Renewable Energy and Climate Change

Renewable Energy and Climate Change

Volker Quaschning
Berlin University of Applied Sciences HTW, Germany

Translated by Herbert Eppel
HE Translations, Leicester, UK
www.HETranslations.uk

Second Edition

Authorised translation of original German text *Erneuerbare Energien und Klimaschutz, 4.A.* © 2018 Carl Hanser Verlag, München. All rights reserved.

Edition History

John Wiley & Sons Ltd (1e, 2010)

Registered Office(s)

John Wiley & Sons, Inc., 111 River Street, Hoboken, NJ 07030, USA

John Wiley & Sons Ltd, The Atrium, Southern Gate, Chichester, West Sussex, PO19 8SQ, UK

Editorial Office

The Atrium, Southern Gate, Chichester, West Sussex, PO19 8SQ, UK

For details of our global editorial offices, customer services, and more information about Wiley products visit us at www.wiley.com.

Wiley also publishes its books in a variety of electronic formats and by print-on-demand. Some content that appears in standard print versions of this book may not be available in other formats.

Library of Congress Cataloging-in-Publication Data:

Names: Quaschning, Volker, 1969– author. | Eppel, Herbert (Translator), translator.

Title: Renewable energy and climate change / Volker Quaschning, Berlin University of Applied Sciences HTW, Germany ; (translator) Herbert Eppel.

Other titles: Erneuerbare Energien und Klimaschutz. English

Description: Second edition. | Chichester, West Sussex, U.K. ; Hoboken, N.J. : John Wiley & Sons Ltd, [2020] | English translation of German text. | Includes bibliographical references and index. |

Identifiers: LCCN 2018052859 (print) | LCCN 2018060578 (ebook) | ISBN 9781119514886 (Adobe PDF) | ISBN 9781119514879 (ePub) | ISBN 9781119514862 | ISBN 9781119514862 (hardcover) | ISBN 111951486X(hardcover)

Subjects: LCSH: Renewable energy sources. | Climatic changes.

Classification: LCC TJ808 (ebook) | LCC TJ808 .Q38913 2020 (print) | DDC 363.738/74–dc23

LC record available at https://lccn.loc.gov/2018052859

Cover Design: Wiley

Cover Images: Top: © BIHAIBO/Getty Images, Bottom: © WangAnQi/Getty Images

Set in 10/12pt WarnockPro by Aptara Inc., New Delhi, India

Contents

Preface to First Edition

The problems of energy and climate change have finally ended up where they belong: at the heart of public attention. Yet the connection between energy use and global warming is something we have been aware of for decades. In the late 1980s the German federal government declared climate change to be one of its main goals. At the time numerous experts were already calling for a speedy restructuring of the entire energy supply. Despite the government's declaration, the official response was, at best, half-hearted. But the climate problem can no longer remain on the back burner. There is a growing awareness that climate change has already begun. The prognosis of the researchers studying what is happening to our climate is extremely serious. If we do not pull the emergency cord soon, the catastrophic consequences of climate change will far exceed even the power of our imagination. The fact that we have awarded the Nobel Peace Prize to Al Gore, the US climate activist, and to the Intergovernmental Panel on Climate Change, both of whom have been urgently warning of the consequences for years, could be seen as a sign of helplessness rather than optimism about our ability to solve the problem.

Just as climate change threatens our environment, new records for rising oil and natural gas prices show that current supplies will not last long enough to cover future demand, therefore, alternatives must be found as soon as possible.

And yet the solution is a simple one: renewable energy. Renewable energy could completely cover all our energy supply needs within a few decades. This is the only way to end our dependence on energy sources like oil and uranium – costly both economically and environmentally – while satisfying our hunger for energy in a sustainable and climate-friendly manner.

However, for many, the route required to reach that goal is still far from clear. Many people still do not believe renewable energy offers a viable option. Some underestimate the alternatives to such an extent that they predict a return to the Stone Age once oil and coal supplies have been fully depleted.

The aim of this book is to eliminate these prejudices. It describes, in a clear and intelligible style, how the different technologies work, which are now available, and the potential for implementing these various forms of renewable energy, with the focus on the interaction between the different technologies. By showcasing some examples of how Germany has tackled this issue, we can show the forms that sustainable energy supply can take and how it can be implemented. However, this book is designed to show all readers, no matter where they live, how they themselves can make a contribution towards building a climate-compatible energy economy. In addition to explaining different energy measures that individuals can undertake for themselves, this book provides concrete planning aids for implementing renewable energy systems.

This book has been consciously written to offer essential facts to a broad spectrum of readers. It introduces the different technologies to anyone new to the subject, while at the same time, providing interesting background information to those with some existing knowledge.

The book has been translated from the original German version. It is an important addition to my technical work 'Renewable Energy Systems', published by Hanser, and also supplements my many lectures on the subject. It is clear from the high level of interest generated by this technical book, now in its sixth edition in German and translated into both English and Arabic, that there is a real need for more literature on the subject of renewable energy. This book should fill this gap and provide support in the development of sustainable energy supply systems.

Berlin, 2009 *Prof. Volker Quaschning*

Preface to the Second Edition

Our excellent sales figures and the positive response to this reference book have shown that this topic and the style in which it is presented appeal to a broad readership. Despite careful checking, minor errors and inconsistencies are sometimes unavoidable. Special thanks are therefore due to all readers who have provided feedback. This second English edition is based on the fourth revised and enlarged German edition. It contains current data on renewable energies and has been expanded to include the latest trends. A dedicated section explains what steps are necessary to comply with the Paris climate protection agreement and thus preserve the livelihoods of future generations. We hope that this book can make a small contribution to accelerating our energy transition at the rate that is required.

Berlin, summer 2018

Prof. Volker Quaschning
HTW Berlin – University of Applied Sciences
www.volker-quaschning.de

1

Our Hunger for Energy

Most people will have heard of the cult TV series, Star Trek. Thanks to this programme, we know that in the not-too-distant future humans will start exploring the infinite expanses of the universe. The energy issue will have been resolved long before then. The Warp drive discovered in 2063 provides unlimited energy that Captain Kirk uses to steer the starship Enterprise at speeds faster than light to new adventures. Energy is available in overabundance; peace and prosperity rule on Earth and environmental problems are a thing of the past. But even this type of energy supply is not totally without its risks. A warp core breach can cause as much damage as a core meltdown in an ancient nuclear power plant. Warp plasma itself is not a totally safe material, as the regular viewers of Star Trek very well know.

Unfortunately – or sometimes fortunately – most science fiction is far removed from the real world. From our perspective the discovery of a warp drive seems highly unlikely, even if dyed-in-the-wool Star Trek fans would like to think otherwise. We are currently not even close to mastering comparatively simple nuclear fusion. Consequently, we must rely on known technology, whatever its drawbacks, to solve our energy problems.

In reality, energy use has always had a noticeable impact on the environment. Looking back today, it is obvious that burning wood was less than ideal, and that the harmful noxious fumes created by such fires considerably reduced the life expectancy of our ancestors. A fast-growing world population, increasing prosperity and the hunger for fuel that has developed as a consequence have led to a rapid rise in the need for energy. Although the resulting environmental problems may only have affected certain regions, the effects of our hunger for energy can now be felt around the world. Overconsumption of energy is the main trigger for the global warming that is now threatening to cause devastation in many areas of the world. However, resignation and fear are the wrong responses to this ever-growing problem. There are alternative energy sources to be tapped. It is possible to develop a long-term safe and affordable energy supply that will have only a minimal and manageable impact on the environment. This book describes the form this energy supply must take and how each individual can contribute towards a collective effort to halt climate change. But first it is important to take a close look at the causes of today's problems.

Renewable Energy and Climate Change, Second Edition. Volker Quaschning.
© 2020 John Wiley & Sons Ltd. Published 2020 by John Wiley & Sons Ltd.

1.1 Energy Supply – Yesterday and Today

1.1.1 From the French Revolution to the Early Twentieth Century

At the time of the French Revolution at the end of the eighteenth century, animal muscle power was the most important source of energy. Around 14 million horses and 24 million cattle with an overall output of around 7.5 billion watts were being used as work animals [Köni99]. This corresponds to the power of more than 100 000 mid-range cars.

> **Power and Energy, or the Other Way Around**
>
> The terms 'power' and 'energy' are closely linked, and for this reason they are often confused with one another and used incorrectly.
> *Energy* is stored work; thus, the possibility to perform work. It is identified by the symbol *E*. The symbol for work is *W*.
> *Power* (symbol: *P*) indicates the time during which the work is to be performed or the energy used.
>
> $$P = \frac{W}{t} \left(\text{power} = \frac{\text{work}}{\text{time}} \right)$$
>
> For example, if a person lifts a bucket of water, this is considered work. The work that is performed increases the potential energy of the bucket of water. If the bucket is lifted up twice as quickly, less time is used and the power is doubled, even if the work is the same.
> The unit for power is the watt (abbreviation: W) (The fact that the abbreviation for watt is the same as the symbol for work does not simplify matters.)
> The unit for energy is watt second (Ws) or joule (J). Other units are also used for energy. Appendix A.1 provides the conversion factors between the different units of energy.
> As the required powers and energies are often very high, prefixes such as mega (M), giga (G), tera (T), peta (P), and exa (E) are frequently used (see Appendix A.1).

The second staple energy source at this time was firewood, which was so important that it probably changed the political face of Europe. It is believed today that the transfer of the Continent's centre of power from the Mediterranean to north of the Alps came about because of the abundance of forests and associated energy potential there. Although the Islamic world was able to maintain its position of power on the Iberian peninsula well into the fifteenth century, one of the reasons why it lost its influence was the lack of wood. The problem was that there was not enough firewood that could be used to melt down metal to produce cannons and other weapons. This goes to show that energy crises are not just a modern phenomenon (Figure 1.1).

In addition to muscle power and firewood, other renewable energies were used intensively until the beginning of the twentieth century. Between 500 000 and 600 000 water mills were in operation in Europe at the end of the eighteenth century. The use of wind power was also widespread, particularly in flat and windy areas. For example, the United Netherlands had around 8000 working windmills at the end of the seventeenth century.

For a long time, fossil energy sources were only of secondary importance. Although coal from underground deposits was known to be a source of energy, it was largely

Figure 1.1 Firewood, working animals, wind and water power supplied most of the energy needed in the world as late as the eighteenth century.

avoided. It was not until a lack of wood in certain areas of Europe led to energy shortages that coal deposits began to be exploited. In addition, the higher energy density of coal proved to be an advantage in the production of steel. In 1800, 60% of coal was used to provide domestic heat, but 40 years later far more coal was used in ironworks and other factories than in homes.

Fossil Energy Sources – Stored Solar Energy

Fossil energy sources are concentrated energy sources that evolved from animal and plant remains over very long periods of time. These sources include oil, gas, hard coal, brown coal, and turf. The base materials for fossil energy sources could only develop because of their conversion through solar radiation over millions of years. In this sense, fossil energy sources are a form of stored solar energy.

From a chemical point of view, fossil energy sources are based on organic carbon compounds. Burnt in conjunction with oxygen, they not only generate energy in the form of heat, but also always produce the greenhouse gas carbon dioxide as well as other exhaust gases.

In around the year 1530, coal mines in Great Britain were producing about 200 000 tons of coal annually. By 1750 it was about 5 million tons, and in 1854 an astonishing 64 million tons. By 1900 three countries, Britain, the USA, and Germany, had an 80% share of world production [Köni99].

ⓘ Renewable Energies – Not That New

The supplies of fossil energies, such as oil, natural gas, and coal, are limited, and they will be depleted within a few decades and cease to exist. Renewable energy sources, on the other hand, 'renew' themselves on their own. For example, if a hydropower plant takes the power of the water from a river, the river will not stop flowing. The energy content of the river renews itself on its own because the sun evaporates the water and the rain feeds the river again.

Renewable energies are also referred to as 'regenerative' or 'alternative' energies. Other renewable energies include wind power, biomass, the natural heat of the earth, and solar energy. Even the sun will eventually disappear in around four billion years. Compared to the few decades that fossil energy sources will still be available to us, this time period seems infinitely long.

Incidentally, renewable energies have been used by mankind for considerably longer than fossil fuels, although the current systems for using these fuels are vastly more advanced than in the past. Therefore, it is not renewable energies that are new, but rather the knowledge that in the long term renewable energies are the only option for a safe and environmentally compatible energy supply.

At the end of the twentieth century, worldwide coal production reached almost four billion tons. With an overall share of less than 3% of the world market, Germany and Britain had lost their former position of supremacy in the coal industry. China and the USA are currently the main coal-producing countries by a considerable margin. Most of the coal produced today is used in power plants.

1.1.2 The Era of Black Gold

Like coal, oil consists of conversion products from animal and plant substances, the biomass of primeval times. Over millions of years plankton and other single-celled organisms were deposited in sea basins. Due to the lack of oxygen, they were unable to decompose. Chemical processes of transformation eventually turned these substances into oil and gas. The biomass that was originally deposited originated from the sun, which means that fossil energy sources like coal, oil, and gas are nothing more than long-term conservers of solar energy. The oldest oil deposits are around 350 million years old. The area around the Persian Gulf where most oil is exploited today was completely below sea level 10–15 million years ago.

The oil deposits were developed much later than coal, because for a long time there were no practical uses for this liquid energy source. Oil was used in small quantities for thousands of years for medicinal and lighting purposes, but its high flammability compared to coal and charcoal gave it the reputation of being a very dangerous fuel. At the end of the nineteenth century petroleum lamps and later the invention of internal combustion engines finally provided a breakthrough.

Industrial oil production began in August 1859, as the American Edwin L. Drake struck oil whilst drilling at a depth of 20 m near Titusville in the US state of Pennsylvania. One name in particular is linked with further oil exploitation in America: John Davison Rockefeller. In 1862 at the age of 23 he founded an oil company that

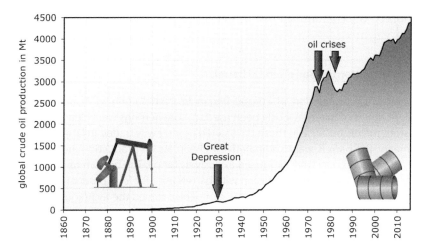

Figure 1.2 Oil production since 1860.

became Standard Oil and later the Exxon Corporation and incorporated large sections of the American oil industry.

However, it was still well into the twentieth century before fossil energy supplies, and specifically oil, dominated the energy market. In 1860 about 100 000 tons of oil were produced worldwide; by 1895 it was already 14 million tons. German government figures reveal that in 1895 there were 18 362 wind engines, 54 529 water engines, 58 530 steam engines and 21 350 internal combustion engines in use in the country [Gasc05]. Half of the drive units were actually still operated using renewable energy sources.

There was a huge rise in oil production in the twentieth century. By 1929 output had already risen to over 200 million tons and in the 1970s it shot up to over three billion tons (Figure 1.2). Today oil is the most important energy source of most industrialized countries. An average German citizen, including infants and pensioners uses 1700 l every year. This amounts to 10 well-filled bathtubs.

Being too dependent on a single energy source can become a serious problem for a society, as history shows. In 1960 OPEC (Organization of Petroleum Exporting Countries) was founded, with headquarters in Vienna. The goal of OPEC is to coordinate and standardize the oil policies of its member states. These include Algeria, Ecuador, Gabon, Indonesia, Iraq, Qatar, Kuwait, Libya, Nigeria, Saudi Arabia, Venezuela, and the United Arab Emirates, who between them at the end of the twentieth century controlled 40% of worldwide oil production. As a result of the Yom Kippur war between Israel, Syria, and Egypt, the OPEC states cut back on production in 1973. This led to the first oil crisis and a drastic rise in oil prices. Triggered by shortfalls in production and uncertainty after the revolution in Iran and the ensuing first Gulf War, the second oil crisis occurred in 1979 with oil prices rising to USD38 per barrel.

The drastic rise in oil prices set back world economic growth and energy use by about four years. The industrialized nations, which had become used to low oil prices, reacted sharply, resulting in schemes such as car-free Sundays and programmes promoting the use of renewable energies. Differences between the individual OPEC states in turn led to a rise in production quotas and a steep drop in price at the end of the 1980s. This also sharply reduced the commitment of the industrialized nations to use renewable energies.

> ### From Alsatian Herring Barrels to Oil Barrels
>
> Commercial oil production in Europe began in Pechelbronn in Alsace (now France) in 1735. Barrels that had previously been used to store herrings were cleaned and then used to store the oil, because in those days salted herring was traded in large quantities, which meant the barrels were comparatively cheap. As oil production increased, special barrels of the same size were produced exclusively for oil. The bottom of the barrels was painted blue to prevent any confusion with barrels used for food products. When commercial oil production began in the USA, the companies copied the techniques used in the Alsace region. This also included the standard size of herring barrels. Since then the herring barrel volume has remained the international measuring unit for oil. The abbreviation of barrel is bbl, which stands for 'blue barrel' and means a barrel with a blue base.
> 1 petroleum barrel (US) = 1 bbl (US) = 158.987 l (litres)

The dramatic collapse in the price of crude oil from almost USD40 a barrel to USD10 created economic problems for some of the production countries, and also made it unattractive to develop new oil sources. In 1998 unity was largely restored again among the OPEC states. They agreed on lower production quotas in order to halt any further drop in prices. In fact, prices rose even higher than originally intended. Now the lack of investment in energy-saving measures was coming home to roost. The economic boom in China and in other countries further boosted the demand for oil to such an extent that it was difficult to meet and, as a consequence, oil prices kept climbing to new record highs. Even though the oil price has fallen sharply again since the financial crisis, new record prices are expected again due to the limited supplies available.

Yet, there have been some fundamental changes since the beginning of the 1980s. In many industrialized countries, energy use has decreased despite rapid and sustained economic growth. The realization has set in that energy use and gross national product are not inextricably linked. It is possible for prosperity to increase even if energy use levels or drops. Nonetheless, the chance to develop true alternatives to oil and to make energy-saving options the norm was missed due to the long period of continuous low oil prices.

This is particularly apparent in the transport sector where cars became faster, more comfortable, heavier, and with more horsepower, but only minimally more fuel-efficient. Today, the fortunate drivers of company cars with 50 hp more than 20 years ago, regularly get stuck in traffic jams (made bearable by air-conditioning and high-tech stereo systems). The tank is also bigger, so that the heavier car, with virtually unchanged consumption, can reach the next petrol station selling cheap fuel. As a result of all the talk about climate change and high oil prices, car manufacturers are now scrambling to incorporate features into their cars that have not been demanded in decades: fuel efficiency and low emissions of greenhouse gases. Since many car companies are struggling with the new requirements, they continue to rely on tried and tested concepts. Because of their political influence, they are able to prevent or dilute the strict savings targets urgently needed for climate protection. Or, like the VW Group, they try to circumvent existing regulations with illegal methods. If Volkswagen had invested the fines paid in the USA in the development of emission-free electric cars, the company would no doubt have been a world leader in this field and would also have made an enormous contribution to climate protection. In retrospect, it is likely the VW scandal will turn out to be a great stroke of luck for Germany. It has shown the technical saving limits of conventional combustion engines and considerably accelerated the switch to electric cars. In the end, it may even have prevented German car makers from falling behind internationally by uncompromisingly sticking to old technologies.

As important as oil is as a fuel, that is not its only use, because it is also an important raw material for the chemical industry. For example, oil is used as a basic material in the production of plastic chairs, plastic bags, nylon tights, polyester shirts, shower gels, scents, and vitamin pills.

1.1.3   Natural Gas – The Newest Fossil Energy Source

Natural gas is considered to be the cleanest fossil energy source. When natural gas is burnt, it produces fewer harmful substances and climate-damaging carbon dioxide than oil or coal. However, this does not change the fact that the combustion of natural gas also produces far too many greenhouse gases for effective climate protection.

The base material for the creation of natural gas was usually green plants in the flat coastal waters of the tropics. The Northern German lowland plains were part of this area 300 million years ago. The lack of oxygen in coastal swampland prevented the organic material from decomposing and so it developed into peat. As time went by, new layers of sand and clay were deposited on the peat, which during the course of millions of years turned into brown and bituminous coal. Natural gas then developed from this due to the high pressure that exists at depths of several kilometres and temperatures of 120–180 °C.

However, natural gas does not consist of a single gas, but rather a mixture of different gases whose composition varies considerably depending on the deposit. The main component is methane, and the gas also often contains relatively large quantities of hydrogen sulphide, which is poisonous and even in very small concentrations smells of rotten eggs. Therefore, natural gas must often first be purified in processing plants using physiochemical processes. As natural gas deposits usually also contain water, the gas must be dried to prevent corrosion in the natural gas pipelines (Figure 1.3).

Natural gas was not seen as a significant energy source until relatively recently. It was not until the early 1960s that natural gas was promoted and marketed in large quantities. The reasons for this late use of natural gas compared to coal and oil is that extracting it requires drilling to depths of several thousand metres. It also requires complicated transport. Whereas oil was initially still being transported in wooden barrels, gas requires pressure storage or pipelines for its transport. Nowadays, pipelines extend for thousands of kilometres from the extraction sites, all the way to providing gas heating to family homes. The world's largest gas producer is Russia, followed by the USA, Canada, Iran, Norway, and Algeria.

However, the demand for natural gas is not constant over the whole year. In countries with cold winters the demand in winter is often double what it is in summer. As it is not economical to cut summer production by a half, enormous storage facilities are needed to balance the uneven seasonal demand. So-called salt caverns and aquifer

Figure 1.3 Left: Building a natural gas pipeline in Eastern Germany. Right: Storage facility for 4.2 billion m³ of natural gas in Rehden, 60 km south of Bremen. Source: Photos: WINGAS GmbH.

reservoirs are used. Caverns are shafts dug in underground salt deposits from where the stored gas can quickly be extracted – for instance, to cover sudden high demand. Underground aquifer reservoirs are suitable for the storage of particularly large quantities of gas. Hence this rock is again filled with what it had stored for over 300 million years and taken from it in a few decades. In total, Germany has a natural gas storage capacity amounting to more than 30 billion cubic metres in operation, in planning or under construction. This corresponds to a cuboid with a base area of 20 by 20 km and a height of 75 m. Environmentally compatible hydrogen is expected to play an important role in future energy supply in the foreseeable future. The existing natural gas storage facilities are already sufficient to compensate for seasonal fluctuations in a completely renewable energy supply. Therefore, natural gas storage facilities and networks will very soon play a central role in securing a sustainable energy supply in the future.

1.1.4 Nuclear Power – Split Energy

In December 1938 Otto Hahn and Fritz Strassmann split a uranium nucleus on a simple laboratory bench at the Kaiser-Wilhelm Institute for Chemistry in Berlin-Dahlem, thereby laying the foundation for the future use of nuclear energy. The laboratory bench can still be admired today at the Deutsches Museum in Munich.

In the experiment a uranium-235 nucleus was bombarded with slow neutrons. The nucleus then split, producing two atomic parts, krypton and barium, as well as two or three other neutrons. With a large quantity of uranium-235, these new neutrons can also split uranium nuclei that in turn release neutrons, thus leading to a chain reaction. If enough uranium is available, the uncontrolled chain reaction will create an atomic bomb. If the speed of the chain reaction can be controlled, uranium-235 can also be used as fuel for power plants.

Germany as an Example of Nuclear History

The Paris Treaty of 5 May 1955 allowed Germany non-military use of nuclear energy. Expectations for the nuclear industry ran high. A separate ministry for nuclear energy was created, and the first minister was Franz Josef Strauss. On 31 October 1957, Germany put its first research reactor, called the nuclear egg, into operation at the Technical University in Munich. In June 1961 the Kahl nuclear power plant fed electricity into the public grid for the first time. In 1972 the Stade and Wuergassen commercial nuclear power plants began to provide electricity, and with Biblis the world's first 1200 MW block went into operation in 1974. In 1989 the last new power plant, Neckarwestheim, was connected to the grid. Until that point the federal government had invested over 19 billion euros in the research and development of nuclear energy. However, public concerns about the risks of nuclear energy continued to grow and prevented the building of new power plants. In 2000, Germany decided to phase out nuclear power. In 2011, another federal government significantly extended the operating times again, but the phase-out was reinstated in the same year, following the accidents at the Fukushima nuclear power plant. The last nuclear power plant in Germany is scheduled to be disconnected from the grid in 2022. Despite more than 50 years of nuclear energy use in Germany, the problem of end storage for highly radioactive waste has still not completely been resolved.

In nuclear fission there is a so-called mass defect, i.e. the total mass of the fission particles is less than that of the original uranium nucleus. A complete fission of 1 kg of uranium-235 produces a mass loss of a single gram. This lost mass is then completely converted into energy. An energy mass of 24 million kilowatt hours is thereby released. Around 3000 tons of coal would have to be burnt to release the same amount of energy.

After Hahn's discovery the use of nuclear energy was promoted mainly by the military. Albert Einstein, who emigrated to the USA in 1933 to escape Nazi persecution, sent a letter to US president Roosevelt on 2 August 1939 warning him that Hitler's Germany was making a serious effort to produce pure uranium-235 that could be used to build an atomic bomb. When the Second World War broke out on 1 September 1939, the American government set up the Manhattan Project with the aim of developing and building an effective atomic bomb.

The biggest problem turned out to be the ability to produce significant quantities of uranium-235 to maintain a chain reaction. If metallic uranium is refined from uranium ore, there is a 99.3% probability that it will consist of heavy uranium-235. This is practically useless for producing a bomb. It even has the characteristic of decelerating and absorbing neutrons, thus bringing any kind of chain reaction to a halt. Only 0.7% of available uranium consists of uranium-235, which must be enriched proportionally higher to create a chain reaction. No separation between uranium-235 and uranium-238 can be achieved by chemical means because chemically both isotopes are totally identical.

Consequently, other solutions had to be sought. Ultimately, this separation succeeded through the use of a centrifuge, because the isotopes have different masses.

The Manhattan Project cost more than USD2 billion between 1939 and 1945. The desired results were finally achieved under the direction of the physicist J. Robert Oppenheimer: on 16 July 1945, two months after the capitulation of Germany, the first test of the atomic bomb was carried out in the US state of New Mexico. Using the bomb on Germany was no longer up for discussion, but shortly before the end of the Second World War the atomic bomb was dropped on the Japanese cities Hiroshima and Nagasaki – with the well-known aftermath.

The non-military use of nuclear energy came some years later. Although physicists like Werner Heisenberg and Enrico Fermi had been conducting tests in reactors since 1941, it was not until December 1951 in Idaho that the research reactor EBR 1 succeeded in generating electric current using nuclear energy.

• www.bund.net/atomkraft	Friends of the Earth Germany info
• www.atomindustrie.de	Satirical site on the use of nuclear energy
• https://pris.iaea.org/pris	IEA Power Reactor Information System
• www.wiseinternational.org	World Information Service on Energy
• www.no2nuclearpower.org.uk	News and information about the UK nuclear industry

Unlike the uncontrollable chain reaction that occurs when an atomic bomb explodes, nuclear fission in a nuclear power plant should occur in a controlled manner. Once a chain reaction has started, the number of new neutrons resulting from the nuclear fission must be kept to a limit. Each splitting of a uranium nucleus releases two to three neutrons, only one of which is allowed to split another nucleus. Control rods that capture the neutrons reduce the number of neutrons released. If this number is too high, the process gets out of control. The nuclear power plant then starts to act like an atomic bomb and an uncontrolled chain reaction occurs. Technically (and this was the prevailing view at the time), nuclear fission can be controlled, and undesired reactions eliminated.

The early euphoria that came with the use of nuclear energy died down when an accident occurred with a reactor on 28 March 1979 in Harrisburg, the capital of the US state of Pennsylvania. Large amounts of radioactivity escaped. Many animals and plants were affected and the number of stillbirths among the nearby population increased dramatically after the tragedy.

On 26 April 1986 another serious accident occurred in a nuclear reactor at Chernobyl, a city in the Ukraine. What was thought to be officially improbable actually happened: the chain reaction got out of control and the result was a nuclear meltdown. The radioactivity that was released produced high radiation levels in places as far away as Germany. Many helpers who tried to contain the damage on site paid for their efforts with their lives and thousands of people died of cancer in the years that followed.

On March 11 2011, the Japanese nuclear power plant Fukushima Daiichi was hit by a strong earthquake and a severe tsunami. The plant was not designed for such an event, and the reactor cooling system failed. As a result, nuclear melts and several explosions occurred, which destroyed four of the six reactors and released considerable amounts of radioactivity. Around 150 000 residents of the area were evacuated and hundreds of thousands of animals, which had been left behind, starved to death.

Another problem with the civilian use of nuclear energy is the disposal of radioactive waste. The use of uranium fuel elements in nuclear power plants produces large quantities of radioactive waste that will create a deadly threat for centuries to come. No safe way has yet been found to dispose of this waste.

Technically, the use of nuclear energy is fascinating and the prospect of generating electricity from relatively small amounts of fuel is very tempting. But there are serious risks involved. Germany has therefore agreed to a general decommissioning of its nuclear energy plants. Once the last nuclear energy plant has been switched off, the country's venture into this field will have cost the German federal government alone more than 40 billion euros in research and development. Germany's most expensive leisure park has become a bizarre showpiece for the incredibly bad investment in nuclear energy. The prototype for a fast breeder reactor was erected at a cost of four billion euros in the North Rhine-Westphalian town of Kalkar. Due to safety concerns, including those relating to the highly reactive cooling agent sodium, the nuclear plant was never put into operation. Today the Kernwasser Wunderland Kalkar leisure park is located in the industrial ruins of the nuclear plant (Figure 1.4).

Conservative politicians and some companies have repeatedly cited nuclear energy as a technology of the future. However, only a small proportion of the many projects announced in recent years have been implemented. Above all, the enormous costs of new nuclear power plants usually quickly end any nuclear dreams. High subsidies are necessary in order to facilitate new nuclear power plants in Europe. For the controversial new Hinkley Point C project in Great Britain, guaranteed electricity prices have been offered that are significantly higher than those for solar and wind power plants. If nuclear energy, which after all is a highly controversial technology, can no longer even offer economic benefits, its days are certainly numbered.

In 2017 there were 449 nuclear power plants in operation worldwide. Yet nuclear energy is relatively unimportant for the global supply of energy. Its share is lower than hydropower and much lower than firewood. If a major effort were made to replace the

Figure 1.4 The Kernwasser Wunderland leisure park is in the grounds of a fast breeder reactor in Kalkar that was never put into operation. Source: Photos: Wunderland Kalkar, www.wunderlandkalkar.eu.

majority of fossil power plants with nuclear energy, uranium supplies would be depleted within a few years. In this sense, nuclear power plants are not a real alternative when it comes to protecting the environment – although this is how some politicians, and specifically the companies that would profit from the use of nuclear power, often like to present the option to the public.

In the long term there are high hopes for a totally new variant of atomic energy: nuclear fusion. The model for this technology is the sun, which releases its energy through a nuclear fusion of hydrogen nuclei. The aim is to duplicate this process on Earth without the danger of triggering an undesirable chain reaction like Chernobyl or Fukushima. But there is a hitch to this plan: the particles must be heated to temperatures of several million degrees centigrade to initiate the momentum of nuclear fusion. There is no known material that can permanently withstand such temperatures. Therefore, other technologies, such as the use of strong magnetic fields to contain reaction materials, are being tested. These technologies have seen some success, but despite the enormous amounts of energy used during the ignition, the reactors always go out by themselves.

Currently no one has seriously predicted whether this technology will ever actually work in practice. Critics point out that the proponents of nuclear fusion have been saying for years that it will take 50 years for a commercial functioning reactor to be connected to the grid. Despite the passage of time that 50-year time frame never reduces – the only thing about nuclear fusion that can be said with any certainty.

However, even if this technology became advanced enough to use, there are two good reasons for opposing the development of nuclear fusion. Firstly, this technology is decidedly more expensive than nuclear fission today. As already mentioned, financing conventional nuclear power plants is difficult today. For economic reasons, preference would be given to alternatives such as renewable energies. Greater investment in fusion testing means less investment in alternative energies. At present we would be happy if we could get a fusion reactor up and running at all. The use of this technology for balancing the grid is therefore hardly conceivable from today's perspective. However, this is precisely what would be needed for fusion power plants to operate in conjunction with renewable power plants, such as solar and wind. Fusion technology is therefore unsuitable for the age of renewables. Not to mention the fact that nuclear fusion plants also produce radioactive substances and waste that present a risk. So, there are very few reasons why government money should continue to be used for this technology.

1.1.5 The Century of Fossil Energy

Whereas traditional renewable energies covered most of the energy needs of mankind until the end of the nineteenth century, the twentieth century can be seen as the century of fossil energy. By the middle of the century fossil fuels in internal combustion engines had almost completely replaced the classic renewable energy systems, such as windmills, water wheels, and vehicles and machines driven by muscle power. Modern hydropower for the generation of electricity and biomass, which was mainly used as fuel, were the only renewable energies of any significance.

After the Second World War the demand for energy soared, and fossil energy sources were able to increase their share substantially. In 2016 fossil energies covered around 79% of the world's primary energy needs (see box p. and Figure 1.5). Hydropower and

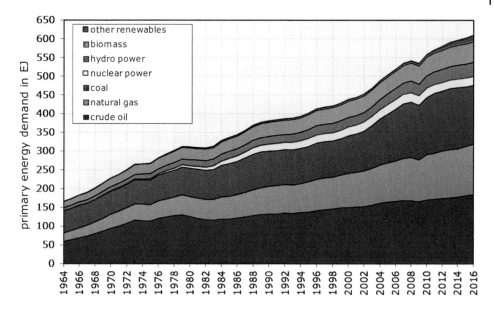

Figure 1.5 Development of primary energy demand worldwide.

nuclear energy had a share of around 6% and 4%, respectively, and biomass around 9%. The other renewable energies amounted to just under 3%. Meanwhile, the situation has started to change. Solar and wind power have continuously high growth rates for new capacity, so their contribution to meeting global energy demand will increase significantly in the coming years. There was a slight decline in the use of coal in 2015 and 2016, while the demand for oil and gas is currently still increasing. If this trend continues, renewable energies could stop the growth in fossil fuels in just a few years and thus initiate effective climate protection measures.

1.1.6   The Renewables Century

Although the share of renewable energies is still comparatively low at present, and the consumption of fossil fuels continues to rise despite all climate protection commitments, the twenty-first century is already on its way to becoming the century of renewables. Many cannot yet imagine a rapid change. Renewable energies share this fate with a multitude of new technologies. Emperor Friedrich Wilhelm II, for example, is said to have initially doubted the change in the transport sector: 'I believe in the horse. The automobile is a temporary phenomenon.'

   The internet and mobile phones have shown us how quickly new technologies can become established. Wind power and photovoltaics are currently expanding rapidly, with growth rates reminiscent of the introduction of the internet and mobile communications. Germany has long been a pioneer in the use of renewable energies. The millionth solar plant was opened there as early as 2011 (Figure 1.6). However, since 2013, when the German government significantly restricted the expansion of renewable energies, other countries, for example China, have taken over Germany's lead role in the expansion of renewable energies. Yet there is no doubt: The age of renewable energies

Figure 1.6 Left: Despite the intensive use of fossil fuels, the expansion of wind energy is booming in the USA. Right: The one-millionth PV installation in Germany. Source: Photos: Dennis Schwartz/REpower Systems SE and BSW-Solar.

has already begun worldwide. Soon they will break the dominance of fossil energies. The only question remains whether the replacement will come in time to stop climate change, which is progressing at an ever-faster pace. Nevertheless, the chances of this happening may be better than many currently dare to hope.

1.2 Energy Needs – Who Needs What, Where, and How Much?

Demand for energy is distributed unevenly across the world. Six countries, namely China, the USA, Russia, India, Japan, and Germany, use more than half the available energy.

The USA alone needs one-sixth of the energy used in the world, even though less than one-twentieth of the worldwide population lives there. If every citizen of India were to use as much energy as the average American, global demand for energy would rise by about 60%. If all the people on Earth developed the same hunger for energy as the USA, demand would increase threefold (Figure 1.7).

> ⓘ **Energy Cannot be Consumed, Actually**
>
> Anyone who has taken physics at high school will have learned about the concept of energy conservation. According to this principle, energy cannot be consumed or produced, but only converted from one form into another.
>
> The car is a good example. The fact that cars consume too much is something we are keenly aware of each time we fill up with petrol. The petrol that a car needs, and we wince every time we pay for it, is a type of stored chemical energy. Combustion produces thermal energy. This is converted by an engine into kinetic energy and transferred to the car. Once

all the petrol has been consumed, the car stops running. However, this does not mean that the energy has disappeared. Instead, it has been dispersed into the environment in the form of waste heat from the engine and heat generated as a result of air resistance and tyre friction. However, this ambient heat is no longer available for practical use; we are unable to produce petrol from ambient heat. When a car is driven, the usable energy content of the petrol is changed into an ambient heat that is no longer usable. This means that this energy is lost to us and thus consumed, even if this is not correct in terms of the laws of physics.

A photovoltaic system is a different matter. It converts sunlight directly into electric energy. It's often said that a solar system produces energy. From the point of view of physics, this is also incorrect. A solar system merely converts hard-to-use solar radiation into high-quality electricity.

When discussing which countries consume the highest amounts of energy, it is important to look beyond overall consumption figures. Population numbers also play an important role in any comparison. In absolute terms, India consumes more energy than Germany or the UK. But this is to be expected with a population of more than one billion people. Consumption per head in India is less than one-sixth of that in Germany or the UK. Although India is the country with the fourth highest use of primary energy in the world, its consumption per head is less than half the world average.

Figure 1.7 shows global primary energy needs per head of population. It is evident that the Western industrialized states and countries with large supplies of crude oil have a high rate of consumption because prosperity and cheap energy prices boost consumption. When it comes to the geographical pattern of consumption, the map clearly shows that the countries with very high consumption – with the exception of Australia, New Zealand, and South Africa – are all in the Northern hemisphere. Germany, France, the

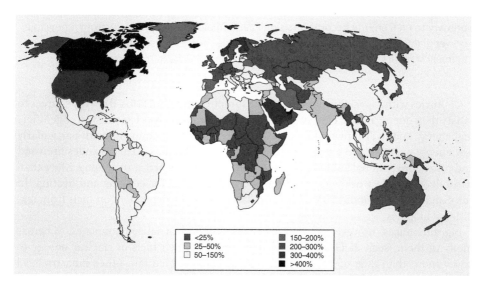

Figure 1.7 Primary energy usage per head related to the world average.

UK, and Italy together consume more than the entire African continent with its population of more than one billion.

Primary Energy, Apple Energy, and Orange Energy

If we compare our own electricity and gas consumption, we see that our consumption will almost always be higher if we heat our homes with gas. A comparison of the gas and electricity bills will not show much of a difference, though. Electricity and natural gas are two types of energy or energy sources that, like apples and oranges, cannot be compared directly like-for-like. Two to three kilowatt hours of gas have to be burnt in a power plant in order to produce 1 kWh of electricity from gas. The rest usually disperses unused into the environment as heat. When comparing different forms of energy, a distinction is therefore made between primary energy, final/secondary energy, and useful energy.

Primary energy is energy in its natural and technically unconverted form, such as coal, crude oil, natural gas, uranium, sunlight, wind, wood, and cow dung (biomass).

Final energy or secondary energy is energy in the form in which it is channelled to users. This includes natural gas, petrol, heating oil, electricity, and district heating (the use of a centralized boiler installation to provide heat for several buildings).

Useful energy is energy in its eventual form, such as light for illumination, warmth for heating and power for machines and vehicles.

The different forms of energy are most frequently compared on a primary energy basis. More than 90% of the original energy content is lost during the conversion of primary energy to usable energy. The classification of renewable energies is not always fully consistent. According to the definition, electricity from solar or wind power plants would be a final energy. However, many statistics refer to it as primary electricity and regard it as primary energy. We can only speculate about the reasons. On the one hand, the statistical identification of the 'real' primary energy relating to renewable electricity is difficult; on the other hand, the renewables resource is so large that efficiencies in the conversion from primary to final energy become less significant. Analogous to electricity from renewable power plants, hydrogen produced from renewable sources is also rated as primary energy in many statistics, although this is also a type of final energy in the narrower sense.

Countries with especially high energy consumption mostly use fossil energy sources to satisfy their energy needs. However, there are exceptions such as Iceland, where geothermal energy and hydropower dominate. On the other hand, countries with particularly low energy needs rely to a large degree on traditional biomass. This includes firewood and other conventional animal or plant products, such as dried animal dung. More than two billion people worldwide use firewood and charcoal for cooking and heating. In sub-Saharan Africa about 90% of the population is totally dependent on fuels from traditional biomass.

Big differences, however, also exist between the industrialized countries. Whereas many of them, such as Germany and the USA, use fossil fuels or nuclear energy to cover more than 80% of their primary energy demand, certain other industrialized countries have already increased their share of renewable energy use considerably. The Alpine countries, Norway, and Sweden use a noticeably high proportion of hydropower.

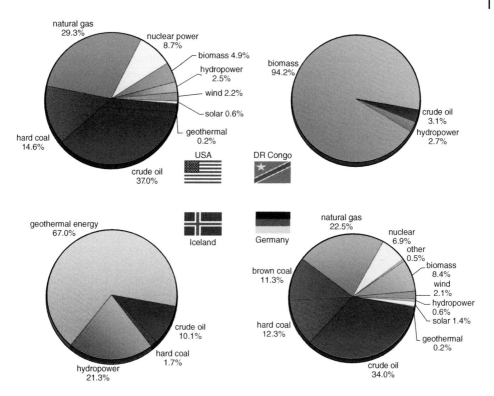

Figure 1.8 Percentage of different energy sources covering primary energy demand in the DR Congo (2015), Germany (2016), Iceland (2015), and the USA (2016).

Biomass also plays a big role in some countries like Sweden and Finland. In Iceland the natural heat of the earth is the energy form with the highest share. Hydropower and geothermal energy together cover well over 80% of Iceland's energy demand.

Ethiopia, on the other hand, is a typical example of one of the poorest countries in the world. More than 90% of the energy it uses is still based on traditional biomass. Figure 1.8 shows the difference in how four countries use key forms of energy to cover their energy needs.

1.3   'Anyway' Energy

According to statistics, only around 1.4% of Germany's primary energy consumption came from solar energy in 2016. In the UK and the USA, the percentage is even lower. The proportion of other renewable energies is also still quite low, making it difficult for most of us to imagine that renewable energies will be riding to the rescue of the environment in a few years. In reality however, renewable energies already constitute over 99% of German energy resources if one looks at the complete picture of energy use.

Winston Churchill supposedly said: 'The only statistics you can trust are the ones you have falsified yourself'. It is widely believed that fossil fuel sources cover the lion's share of our energy needs. At least this is what all the usual statistics on energy claim. But it is only true if we define our energy demand in a very narrow way.

The heat of a radiator, the light provided by a conventional light bulb, and the driving energy of a ship's diesel engine generally form an integral part of our energy demand. What is not included in any statistics on energy is the warming effect of the sunshine streaming through windows, the sunlight that illuminates houses and streets so that artificial lighting can be switched off during daylight, and the wind that can propel sailing boats right across the Atlantic. A heated greenhouse that uses artificial light to grow useful plants is included in the statistics on energy; on the other hand, a covered early planting of vegetables that uses only natural sunlight is not included. The floodlight illumination of a stadium during an evening football game falls under our energy needs. If the football game takes place in the bright sunlight, the statistics on energy will claim that the football arena that is brightly lit up by the sun actually does not need any light. If we switch on snow blowers to compensate for the ever-decreasing amount of snow available in ski areas, this becomes a case for the statistics, whereas natural snow is not. When we fill our drinking water storage containers using electric pumps, we have to pay for the energy used. If rain fills the storage containers, this is not considered in the statistics. The high amount of electricity needed to run electric dryers also increases energy use. On the other hand, if the washing is dried by the wind and the sun on a conventional clothesline, this does not constitute an energy need as far as the statistics are concerned.

Natural and technically unconverted forms of energy are not a component in our energy demand in a conventional sense. Yet it should not make any difference where we derive the energy needed to heat our bath water, grow our plants, or provide light. We take the availability of natural renewable energy forms such as solar energy so much for granted simply because they are there anyway and thus, appear to have so little value that they do not even merit a mention in the statistics. However, this distorts our impression of our energy demand and puts the possibilities of renewable energy in a false light.

This can be illustrated using the example of energy consumption in Germany. Germany covers an area of $357\,093\,km^2$ and the annual solar radiation is on average $1064\,kWh\,m^{-2}$. Germany therefore benefits from 380 trillion kilowatt hours of energy from the sun each year. This is about 100 times as much as the primary energy consumption recorded in the statistics for Germany, and even more than the entire primary energy needs of the world. Part of this radiation heats the earth and the air; another part is converted into plant growth, thus producing biomass.

Around 800 mm or $0.8\,m^3$ of rain fall per square metre in Germany. The annual rainfall for all of Germany adds up to 286 billion cubic metres. The sun evaporates this water before it reaches the Earth in the form of rain. One cubic metre of water requires 627 kWh to evaporate. This means the annual rainfall contains around 170 trillion kilowatt hours of energy.

About 2% of solar energy is converted through the movement of the wind. In Germany this amounts to around eight trillion kilowatt hours. Sun, wind, and water together produce abound 567 trillion kilowatt hours of energy in Germany each year. Geothermal and ocean energy are not even included in these figures. If this quantity of energy were to drop by a small percentage, the result would be drought and arctic winters (Figure 1.8).

The statistics of 2016 listed Germany's primary energy demand as being around 13 petajoules. This converts to close to just under four trillion kilowatt hours. Of course, the statistic includes solar energy, hydropower, and wind energy. The proportion of all renewable energies combined in the primary energy demand totals 0.5 trillion kilowatt

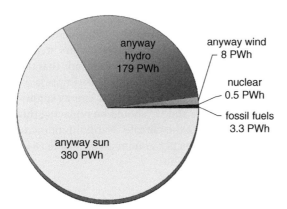

Figure 1.9 Total energy resources in Germany taking into account 'anyway' energy; that is, natural renewable forms of energy.

hours. This is the amount that technical installations convert into renewable energy. The natural forms of renewable energy that exist anyway are totally omitted from this statistic. This explains the small obvious statistical discrepancy for the previous calculation of 567 trillion kilowatt hours for renewable energy resources. The difference between this and conventional statistics is clarified in Figure 1.9, where the natural renewable forms of energy that previously were not recorded in the statistics are referred to as 'anyway' energy – in other words, energy that exists anyway.

Anyone who thinks that these calculations amount to statistical hair-splitting is wrong. As the facts about climate change have now become public knowledge, there is a general interest in replacing fossil fuels with renewable energies as quickly as possible. But many people are under the impression that this is difficult to implement and almost impossible to accomplish within a reasonable period of time. The claim that solar energy only constitutes an insignificant share of energy resources is repeated like a prayer wheel. If this claim were true, this scepticism would be justified. In fact, it is fossil and nuclear energies that make up 0.6% of the energy resources in Germany. And no one could seriously doubt that 0.6% is replaceable in the foreseeable future.

The eruption of the Tambora volcano in Indonesia in 1815 shows us what happens when even a fraction of 'anyway energy' is lost. The gigantic quantities of volcanic gases and dust that were emitted into the atmosphere reduced the amount of sunlight available in the following years. In 1816 and 1817 Europe experienced massive crop failures. Ten thousand people died of starvation. If something like this happened today, we would suffer similar consequences. A large proportion of the energy that safeguards our food supply comes from a natural source – the sun. The small and diminishing supply of fossil and nuclear energy could not come anywhere close to compensating for even relatively minor fluctuations in natural energy.

The question then is what is 'anyway energy' worth? In Europe in 2017 oil cost more around 3 cents (euro) per kilowatt hour before tax; natural gas was around 1.7 cents, and rising. Because solar radiation and wind power cannot be stored as easily as oil and natural gas, their value is assumed to be less than that of natural gas, say 1 cent per kilowatt hour. Hydropower, on the other hand, could be set at 1.5 cents per kilowatt hour because it is easier to store. The total value of 'anyway energy' works out as around 6.5 trillion euros per year. According to this calculation, 'anyway solar energy' alone is worth around 3.8 trillion euros. In the USA, it is worth the equivalent of 100 trillion euros.

Natural renewable forms of energy in the order of 567 trillion kilowatt hours are not recorded as a separate entry in statistical calculations in Germany. This means that the public perception of the energy supply is distorted. We are left with the false impression that fossil and nuclear energy sources make up the major portion of our energy supply. In reality, the share from these sources is less than 1% and we should be replacing this with renewable energies as soon as possible to protect the climate. Natural renewable forms of energy to the value of around 6.5 trillion euros per annum are available to us today free of charge. We cannot afford to ignore this option.

1.4   Energy Reserves – Wealth for a Time

When we use fossil energy sources today, we are utilizing solar energy that was stored millions of years ago – but without the possibility of renewing this source in the foreseeable future. Yet our current hunger for energy is so great that most of the known fossil deposits will be used up during the twenty-first century. Also, suitable deposits of uranium fuel for conventional nuclear plants are becoming rarer.

Conventional or Non-Conventional, that is the Question

No two oil fields on Earth are alike. Some oil deposits are stored in liquid form only at a depth of 100 m below the ground. Others lie at a depth of 10 000 m or are mixed with sand and can only be mined at very high costs, if at all. In order to obtain a better overview of possible ranges, a distinction is therefore made between reserves and resources as well as conventional and unconventional deposits when indicating remaining reserves of crude oil, natural gas, coal, and uranium.

Reserves are proven energy feedstocks that can be economically extracted with today's technology at current prices.

Resources are proven but currently technically and/or economically not recoverable and not proven but presumed, i.e. purely speculative quantities of energy feedstocks. Only a fraction of the resources will therefore be accessible. If technology develops further or raw material prices rise, some resources are gradually added to reserves. Reserves then increase and resources decrease.

Total potential is the sum of reserves and resources. As things stand today, it will not be possible to fully exploit this potential. However, it is possible that new, unexpected deposits may still be found, thus increasing reserves or resources and thus the total potential.

Conventional deposits are reserves or resources that can be developed with conventional production methods. These include crude oil or natural gas in underground cavities that can be extracted via a simple well.

Non-conventional deposits are reserves or resources that are accessible with complex and innovative extraction methods. These include oil sands, oil shale, bitumen or crude oil, and natural gas in smaller cavities in impermeable layers, which first have to be broken up using a so-called fracking process. The extraction of unconventional deposits is often significantly more expensive than conventional ones.

For decades pessimists have been warning about the imminent end to fossil energy reserves. Yet, this end never quite seems to be in sight, and most people take no heed of the warnings. It was not until oil prices started to rise again in 2000 that the message that 'black gold' would one day run out finally sank home.

But the number of new finds, specifically of oil, has declined substantially during recent years, and new supplies cannot be exploited fast enough to meet rising demand. In the long term, oil prices will therefore continue to rise, even if brief dips in prices seem to signal an easing of the situation. On one hand, demand is rising, whereas supplies tend to be dwindling; and on the other hand, the effort needed to exploit new supplies is increasing along with the costs.

During the first commercial drilling in America in 1859, oil could be found at depths of 20 m, whereas today drilling at depths of up to 10 000 m is quite common. Significant technical progress has also been made in locating possible deposits, and so we know far more today about possible finds than we did several decades ago. However, this also makes it highly unlikely that any major new finds will be discovered.

1.4.1   Non-Conventional Reserves – Prolongation of the Oil Age

The strong increase in oil and gas prices since the 1990s has led to interest in developing completely new deposits that are mined using unconventional methods. A veritable new oil and gas fever has broken out in North America. In just a few years' time, the continent on the other side of the Atlantic could even briefly overtake the Middle East in oil and gas production.

In Venezuela, and in the Canadian province of Alberta, enormous quantities of oil sands have been found. These are retrieved in opencast mines. The Canadian mining region covers a gigantic area of $149\,000\,km^2$. This corresponds to approximately the size of England. The clearing of forests alone releases enormous volumes of carbon dioxide. The oil is then separated from the sand using large amounts of water and energy. What remains is heavily polluted wastewater and a devastated landscape. Due to the high energy consumption in oil sands production, carbon dioxide emissions continue to increase. Taking into account the release of greenhouse gases from deforestation, Canada's greenhouse gas emissions rose by an 'impressive' 46% between 1990 and 2010. The end of the oil age is already beginning to leave its dirtiest traces.

In the USA, the era of oil production was already largely over. Most of the exploitable conventional deposits have been exploited. New deposits in Alaska or the deep sea could only be developed to a very limited extent due to the enormous risks for the local environment. The strong military commitment of the USA in the Middle East in recent years has been due in no small measure to securing access to the energy feedstocks there.

But now a new technology has revolutionized the extraction of crude oil and natural gas in the USA: so-called fracking. First, a deep bore opens up the bedrock. Then a liquid is pressed into the well at high pressure, which causes cracks in the rock at depth. The idea is that gas or oil trapped in the rock will escape through these cracks and reach the surface via the borehole. If pure water were used for the process, the cracks would close immediately when the water is pumped back. This is why the water is mixed with sand and numerous chemicals, some of which are very toxic. This is intended to keep the cracks open, facilitate the outflow of oil or gas and prevent the growth of bacteria.

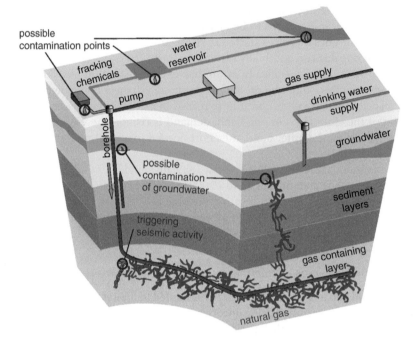

Figure 1.10 Operating principle and risks of natural gas fracking.

Environmentalists criticize numerous incalculable risks associated with fracking (Figure 1.10). Blasting the cracks in the ground can cause small earthquakes. Fracking chemicals can be released into the environment if handled improperly. The disposal of the large quantities of contaminated wastewater that is pumped back is problematic, and chemicals or natural gas can contaminate the groundwater and thus ultimately the drinking water via cracks and gaps. In the USA, drinking water has already been so polluted in various places, thus allowing this water to be ignited by the flammable methane dissolved in it. If natural gas escapes unused into the atmosphere, it also increases the greenhouse effect. The question remains whether a fairly short extension of the oil and gas age justifies such environmental impacts.

1.4.2 An End in Sight

In the past, constant technological advances in the exploitation of oil and natural gas have always resulted in a revision of the forecasts about the lifetimes of reserves. The large coal reserves still available worldwide in particular, could enable us to use fossil energy sources for decades or even another century.

At the current rate of production, the known supplies in the USA, Europe, and Asia will soon be depleted. This increases the dependency, particularly of the industrialized nations, on a small number of producing countries. The USA is trying to reduce this dependence by exploiting unconventional deposits. China is securing more and more access to deposits in other regions such as Africa. More than 60% of extractable oil supplies are found in the Middle East (Figure 1.11). The biggest oil producers in the region are Iraq, Iran, Kuwait, Saudi Arabia, and the United Arab Emirates.

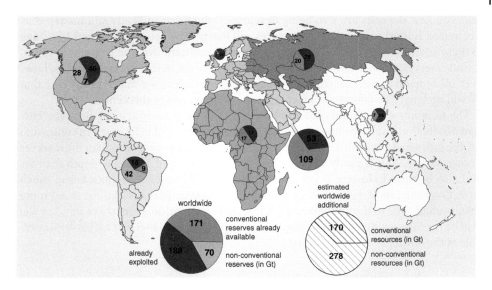

Figure 1.11 Distribution of oil reserves on earth by region (2016). Source: BGR [BGR17].

This region has been the scene of major conflicts in recent years, and its large oil reserves are likely to increase tensions even further in the future. The dependency of the industrialized nations on the OPEC countries will also increase because these countries have almost three-quarters of the known reserves.

The current extent of availability can be calculated by dividing the known exploitable reserves by current production. In the case of oil, this is about 39 years (cf. Figure 1.12). The non-conventional reserves can extend the range by just 16 years. This range could drop further if there is an increase in annual production. In addition to the known

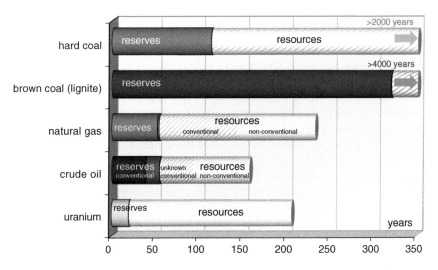

Figure 1.12 Range (in years) of known energy reserves and resources based on current production. Source: Data: BGR [BGR17].

reserves, new deposits are also being developed, which are currently still managed as a resource. It is estimated that the reserves will increase by between 50% and 100% due to these additional deposits. If production remains constant over the next few decades, the oil reserves will last another 100 years or so. However, in just the last 50 years, however, oil demand has almost tripled worldwide. Nevertheless, we can be pretty sure that oil will no longer be relevant as an energy source at the end of this century.

The situation with natural gas and coal supplies is not quite so critical. Based on current production levels, the known gas reserves will be depleted in 52 years. In contrast to oil, the estimated additional resources are considerably more extensive than the reserves known to date. This is partly because the deposits are located at a lower depth than oil, and also because industrial production and the search for new supplies began much later. Over the last 50 years, however, the demand for natural gas has increased more than fivefold. Due to the continuing high level of consumption, natural gas supplies will also be running out during this century. Coal is the only fossil fuel that may still be available at the dawn of the next century.

1.4.3   The End of Fission

A key point about fuel supply, and one that most people are unaware of, is that even uranium supplies are very limited. Although there is more uranium in the Earth's crust than either gold or silver, less than 1% of the purest natural uranium can be used to create energy. Power plants can only use natural uranium after the usable part of uranium has been enriched with uranium U-235.

The share of ore must be higher than average to enable an effective exploitation of natural uranium. Canada is the only country that has deposits with a uranium ore content of more than 1%. If the uranium ore content drops, considerably larger amounts of it need to be mined for its degradation. This substantially increases the energy required to extract the ore and contributes to the costs.

Despite intensive efforts by some countries to develop nuclear energy, at 4%, its share of worldwide primary energy supply is still relatively low. A few countries like France are using nuclear power plants to cover up to 80% of their electricity demand. However, even in France cars cannot be run on nuclear power, and only some houses use nuclear energy for heating. Consequently, nuclear energy only really constitutes around 40% of the total primary energy supply, even in France.

The uranium deposits would be depleted in just a few years if nuclear energy were used to replace all fossil energies. Power plants could use other technologies, such as the risky fast breeder technology, to increase the amount of energy they are able to exploit, but this would do very little to change the fact that the supply of uranium is limited. The fact is that the uranium deposits that can be exploited economically will run out in a few decades at the most – which does not make a convincing argument for building new nuclear power plants with a lifespan of 30–40 years. For these reasons alone, nuclear energy is not a viable alternative to fossil energies.

1.5   High Energy Prices – the Key to Climate Protection

The low energy prices of the 1970s were the foundation not only of the economic miracle in several industrial countries, but of the massive rise in energy consumption. The founding of OPEC and the politically motivated limiting of production in 1973 led ultimately to a dramatic increase in oil prices. The industrialized countries were stunned and reacted with relative helplessness. In 1973 they founded the International Energy Agency (IEA) to coordinate their energy policies and ensure that the supply of energy remained secure and affordable.

The purpose of strategic oil reserves is to guarantee availability when the oil supply is interrupted and to stabilize prices. For example, Germany stockpiles 25 million tons of crude oil or crude oil products that can cover the country's oil demand for around 90 days.

The US strategic petroleum reserve is the largest in the world and holds up to 99 million tons. The commitment to developing the use of renewable energies also increased in the 1970s. However, a large number of failed mammoth projects showed that cost-effective and sustainable energy supply is not something that can be forced through; it can only be the outcome of long-term, ongoing development. Nevertheless, the oil crisis in the 1970s paved the way for the current boom in renewable energies.

The 1990s were marked by extremely low oil prices. As a result, efforts to save energy and to develop renewable energies stagnated. Due to booming global economic activity and extremely high demand, especially from China, oil prices reached new heights after the year 2000. The lack of commitment to reduce oil consumption took its revenge. Oil prices in 2012 were almost double what they were at the time of the oil crises in the 1970s (cf. Figure 1.13). Up to now this has only had a limited effect on the world economy. This can be explained by considering the inflation-adjusted oil prices. In 1980 one US dollar bought around three times as much as it could buy in 2017. The inflation-adjusted oil price at the time was therefore three times as high. Another reason is that the economy today depends considerably less on energy prices than at the time of the oil crisis. However, increasing oil prices would have significant impact on the world economy.

As the supplies of fossil energies begin to run out, oil, natural gas, and coal prices will rise further. The dip in prices since 2015 due to an international price war offers a short

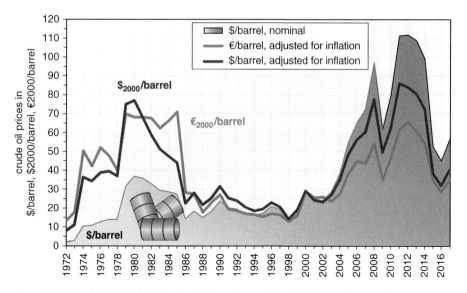

Figure 1.13 Development of oil prices with current prices and inflation-adjusted prices.

breather at best. Another round of price increases is a certainty. Political risks and a growing reliance on certain countries rich in raw materials also conceal the considerable possibility of another sudden hike in prices. For economic reasons it is important that some urgency be given to developing an alternative energy supply beyond fossil or nuclear energy sources.

During the period of transition, supply, and demand will also allow a fluctuation in the price of renewable energies, as was shown by the price increase for wood-burning fuels in 2006. However, in the long term the prices for renewable energies will continue to drop as a result of ongoing technical advances and more efficient production, whereas the price of fossil energy sources and nuclear energy will continue to rise.

Renewable energies are already fully competitive with fossil alternatives in many areas. In 2017, solar plants were already able to undercut prices of 2 cents per kWh^{-1} in tenders in the sunny regions of the world and are thus well below those of new fossil-fuel power plants. Even in less sunny Germany, solar plants can now undercut new fossil-fuel power plants, as there is practically no difference in the price of electricity from new solar and wind power plants at the current stock exchange electricity price.

However, the fossil age is massively supported by subsidies. Worldwide, the unimaginable sum of USD5.3 trillion was spent on subsidies for oil, gas, and coal in 2015. Without this support, the fossil age would soon be over for economic reasons alone. However, further increases in energy prices will sooner or later make this subsidy insanely unaffordable and thus initiate a rapid change towards a sustainable energy supply. Countries that started the transformation process early on will find it easier to master the challenges of the transition.

2

The Climate Before the Collapse

We have known for a long time that the climate is changing. Numerous ice ages and warming periods have shown that the Earth's climate is constantly undergoing change. In terms of human lifetimes, each of these periods lasted a relatively long time. In the more recent history of the Earth, ice ages have occurred about every 100 000 years, always interrupted by much shorter warmer periods, referred to as interglacial periods. Our current warm period, called the Holocene era, began about 11 700 years ago. As the previous interglacial periods on average lasted only about 15 000 years, we should be heading towards the next ice age.

The exact causes for the alternation between interglacial periods and ice ages can only be reconstructed to a point. Natural effects, such as changes in solar activity, changes in the geometry of the Earth's orbit, vulcanism, changes in the ocean currents and the shifting of continental plates are considered the main causes of climate change. When multiple causes occur at the same time, very abrupt changes are possible. This is confirmed by the climate history of the Earth. In this sense, the global warming that we have observed in recent years is nothing out of the ordinary. What is unusual, however, is that, for the first time it is apparently living things on Earth – namely, human beings – that are causing an abrupt change in climate.

2.1 It Is Getting Warm – Climate Changes Today

2.1.1 Accelerated Ice Melt

After the last ice age, global temperatures increased by around 3.5 °C. Due to the warming and thawing ice, sea levels rose by more than 120 m. During the last ice age, areas that are densely populated today were covered by ice sheets several metres thick, and landscapes which were once fertile have been swallowed by the sea. Over the last 7000 years, however, the climate conditions on Earth have been very constant. Sea levels have hardly changed at all and temperatures have only changed by a few tenths of a degree Celsius. This climate stability was one of the essential prerequisites for humanity's development. Our civilization, with its settlements and agricultural areas has adapted to the stable conditions. If we destroy this stability, it will profoundly impact life as we know it today.

A look at developments since the last ice age is also helpful when assessing future temperature changes. Figure 2.1 shows that even relatively small temperature changes can have major effects. A warming of 1 °C does not sound very dramatic to many people

Renewable Energy and Climate Change, Second Edition. Volker Quaschning.
© 2020 John Wiley & Sons Ltd. Published 2020 by John Wiley & Sons Ltd.

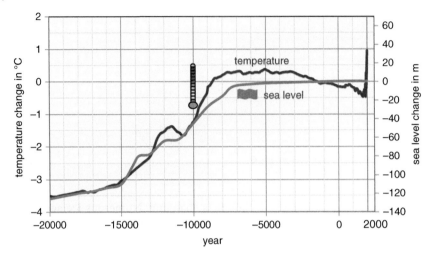

Figure 2.1 Changes in temperature and sea level between 20 000 BCE and 2016 CE. Source: Data: [CDI16, Mar13, Sha12, Fle98], period 1951–1980 corresponds to zero.

at first. However, if this is viewed in relation to the temperature increase since the last ice age, even such a seemingly small value is rather disturbing.

Due to the influence of humans, the temperature has already risen by around 1 °C in the last 100 years, and the rise is accelerating more and more. You don't need to be a climate expert to realize that the recent increase cannot be a normal development. Climate experts are particularly concerned about the comparatively fast rate at which temperatures are rising. There is no natural explanation for this extreme increase.

If further major changes to climatic conditions occur, they will undoubtedly have an even stronger impact on the face of the Earth and our current way of life than any of the most dramatic historical events of the last few centuries have had. Experts therefore consider warming above 1.5–2 °C to be unacceptable. Therefore, today, combatting the greenhouse effect caused by humans and its associated warming is probably, by far, the most important task in preserving the lives of future generations.

- The global surface temperature in 2016 was already 0.94 °C above the average between 1951 and 1980.
- The 2000s were the warmest decade since temperature measurements began.
- The increase in temperature during the last 50 years was twice as high as during the last 100 years. The warming of the Arctic has been twice as fast.
- The temperatures of the last 50 years have been higher than at any time in the last 1300 years.
- Glaciers are shrinking worldwide, as are the ice sheets on Greenland and Antarctica. The Alpine glaciers have already lost two thirds of their volume between 1850 and 2010.
- Summer Arctic sea ice cover decreased from 7.5 million square kilometres in 1982 to 3.5 million square kilometres in 2012.

- Since 1993 the sea level has risen an average of 3.1 mm yr^{-1}; during the twentieth century this amounted to a total of 17 cm. More than half of this is due to the thermal expansion of the oceans, about 25% to the melting of mountain glaciers and around 15% to the melting of the arctic ice sheets.
- The frequency of heavy precipitation has increased.
- The frequency and intensity of droughts have increased since the 1970s.
- The frequency of extreme temperatures has increased.
- Tropical cyclones have become much more intense since the 1970s.

Global warming is not uniform across the world. In the Arctic region, in particular, the temperature change has already exceeded 2 °C in places (Figure 2.2). Generally, the land warms up faster than the oceans. With an average temperature rise of more than 4 °C, some regions could develop literally into death zones where, due to the enormous heat, humans could not survive for long without technical aids.

The temperatures in the polar regions are rising even more quickly than in the rest of the world. The ice coverage in the Arctic has decreased by about 10–15% within 20 years (Figure 2.2). Ice cover in the Arctic has fallen by more than 50% within 30 years (Figure 2.3). Besides the ice masses of the Arctic, many glaciers are melting rapidly. The largest glaciers in the world, the Bering Glacier in the Arctic region of Canada, has shrunk by more than 10 km over the last century. Less than half the mass of the mountain glaciers in the Eastern Alps that existed in 1850 still remains today.

So far, sea levels have only risen by around 20 cm in the last 100 years. However, should the continental ice on Greenland or Antarctica noticeably melt in the future, the rise in sea levels would accelerate noticeably.

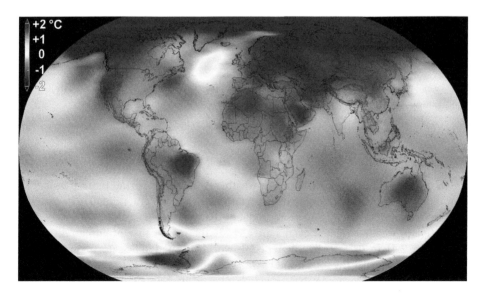

Figure 2.2 Temperature change between 2012 and 2016 compared to the long-standing average between 1951 and 1980. Source: NASA/Goddard Space Flight Center Scientific Visualization Studio, http://svs.gsfc.nasa.gov.

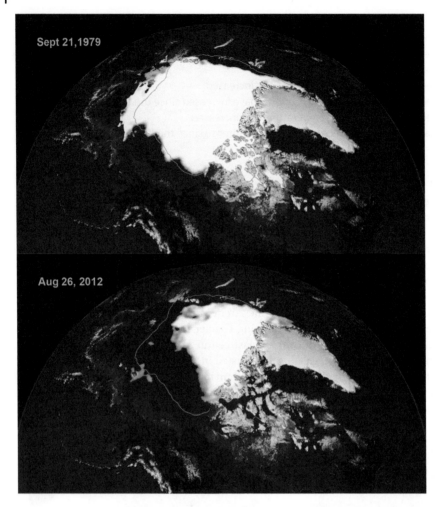

Figure 2.3 Arctic summer ice coverage in 1979 (above) and 2012 (below). Source: NASA, http://svs.gsfc.nasa.gov.

2.1.2 More Frequent Natural Catastrophes

As global temperatures rise, extremes in weather are also occurring more frequently. Major differences in temperatures cause storms to be more violent, rainfall to be heavier and high tides and flooding to be more frequent than before.

Climate and weather-related events are already the main cause of human displacement (Figure 2.4). In 2016, 23.5 million people worldwide had to flee storms and floods. Between 2008 and 2016, the total number was 196 million [iDMC17]. A large proportion of the population displacement has already happened in Asia, Latin America, and the Caribbean, but this has now developed into a problem in Europe also and continues to feature in international news. However, Europe is unlikely to be spared from such movements in the face of rising climate change impacts. If global warming causes sea levels to

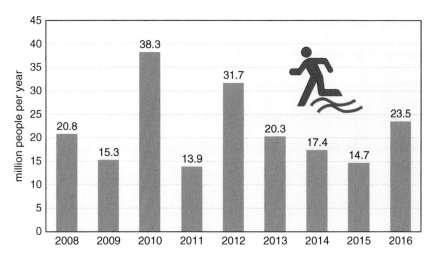

Figure 2.4 Number of people displaced worldwide by climate- and weather-related natural disasters such as storms and floods. Data: iDMC [iDMC17].

rise by only 1 m in the medium term, around 100 million people will permanently lose their homes.

- Winter 1990: The storms Daria, Herta, Vivian, and Wiebke kill 272 people in Europe and cause €12.8 billion in damage.
- 29 April 1991: A storm tide resulting from the tropical cyclone Gorky hits Bangladesh. 138 000 people die. The material damage at €3 billion is comparatively high in this poor country.
- 12 December 1999: Storm Lothar devastates large areas of Europe. As many as 110 people die. The damage amounts to €11.5 billion.
- August 2002: Unusually heavy rainfall of up to 400 l m^{-2} causes major flooding in Germany and some neighbouring countries. Overall in Europe 230 people lose their lives and the damage is €18.5 billion.
- August 2003: Europe's most extreme heatwave in recorded history takes 70 000 lives and causes damage close to €13 billion.
- August 2005: Hurricane Katrina wreaks havoc in the USA and destroys the city of New Orleans, resulting in the deaths of 1322 people. The most expensive storm of all time causes USD125 billion (around €95 billion) worth of damage.
- 18 January 2007: Hurricane Kyrill sweeps across Europe. Deutsche Bahn halts all train travel in Germany for the first time in history.
- October 2010: An unusual drought in East Africa causes dramatic crop failures. Around 260 000 people starve to death.
- October 2012: Hurricane Sandy devastates parts of the Caribbean and the US East Coast and also moves unusually far north, hitting New York hard. A total of 253 people die. The losses amount to USD66 billion (around €50 billion).

- June 2013: Eleven years after the flood of 2002, extreme precipitation once again causes massive flooding and record high water levels in Germany, Austria, and the Czech Republic. Again, there are fatalities and losses in the billions.
- July 2016: Extreme precipitation and flooding cause damage worth USD20 billion in China. 60 million people are affected, 237 die.
- September 2017: Hurricanes Harvey, Irma, and Maria destroy parts of the Caribbean and the US metropolis Houston. The losses are estimated at USD215 billion. 324 people die.
- January 2018: Hurricane Friederike, with wind speeds of up to 200 km h^{-1}, claims 8 lives in Germany.

According to observations by Munich Reinsurance Company, the number of property losses is also increasing continuously, with the global total already exceeding €200 billion in record years.

Hurricane Katrina alone, which devastated the US city of New Orleans in 2005, caused damage worth around USD 125 billion and cost the lives of 1300 people (see also Figure 2.5). In 2017 Hurricane Harvey destroyed large parts of Houston. In some places, within a few days, more than 1500 l of rain per square metre had fallen and 200 000 houses were damaged or destroyed. The overall damage is estimated at USD85 billion.

In Germany, too, extreme events have increased. Examples in recent years have been heavy rain and floods (Figure 2.6). Many people remember the record summer heat of 2003. Major heatwaves cause a decline in harvest yields. Extreme weather events also bring about a rise in the death rate. In the summer of 2003, some 70 000 more people died in Europe than in a normal year as a result of the intense heat. An estimated 7000 heat-related deaths occurred in Germany alone.

Although the financial damage resulting from natural catastrophes is still at present manageable in Europe, these figures are expected to rise dramatically by the end of the century. The German Institute for Economic Research DIW projects that the overall cost of climate change to Germany alone will amount to around 3000 billion euros by the year 2100, if global warming brings about a 4.5 °C rise in temperature [Kem07].

Figure 2.5 Damage caused by hurricanes in the USA. Photos: US Department of Defense|Pixabay.

Figure 2.6 Damage caused by flooding and thunderstorms in Germany. Photos: Wikimedia Commons – Stefan Penninger|Pixabay.

2.2   The Guilty Parties – Causes of Climate Change

2.2.1   The Greenhouse Effect

Without the protective influence of the atmosphere, the Earth's temperature would be around −18 °C. This means we would be living on an ice-bound planet (Figure 2.7).

Various natural trace gases in the atmosphere – such as water vapour, carbon dioxide, and ozone – prevent the Earth from emitting all incoming solar energy back into space. As in a greenhouse, these gases radiate part of this energy back to Earth. This natural greenhouse effect is the basis for life on Earth and as a result, the average temperature has settled at around +15 °C.

Many different possible causes for the observed climate change have already been discussed. For a long time, sceptics even questioned whether climate change was really

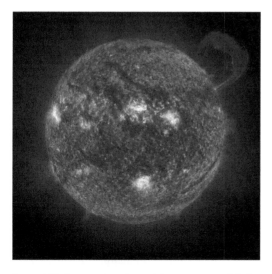

Figure 2.7 Changes in solar activity are responsible for only a fraction of global warming. Picture: NASA.

taking place. Now that no one can seriously claim that it is not getting warmer, there are still some individuals trying to blame it on natural causes – for example, on solar activity. Apparently solar activity during the last few decades has been higher than in all the previous 8000 years. It has been demonstrated that the amount of radiation reaching the Earth has indeed risen slightly. However, scientists rule out the possibility that this is what is causing the high levels of warming today. At most, one-tenth of the observed increase in temperature is attributed to the rise in solar activity.

During the last few millennia, a balance has formed in the level of trace gases in the atmosphere, which has enabled life in the form that we know it today. The most plausible cause of warming is that the proportions of the trace gases have changed significantly due to human influences. The concentration of gases that have been proven to cause global warming has increased considerably in recent decades. Therefore, humans are causing an increase in the natural greenhouse effect. The greenhouse effect brought about by humans is also referred to as an anthropogenic greenhouse effect (Figure 2.10). But this is not a particularly new theory.

Are We Ruining the Air that We Breathe?

The air that we breathe out contains around 4% carbon dioxide – about 100 times more than the air we breathe in. Every year, we each release around 350 kg of carbon dioxide into the atmosphere. When we light a campfire using wood, we are also guilty of releasing carbon dioxide. However, plants, animals, and people are all part of a biogeochemical cycle. People consume carbohydrates and breathe in oxygen and together these are converted into carbon dioxide, which they then breathe out again.

Plants in turn intake this carbon dioxide and then provide us with carbohydrates. Carbohydrates are organic compounds consisting of carbon, hydrogen, and oxygen and are produced in plants through photosynthesis. For example, grains and noodles comprise 75% carbohydrates. It is even possible that the wheat in Italian spaghetti has converted the carbon dioxide that we breathed out during our last Tuscan holiday into carbohydrates.

When a plant burns, rots, or ends up as a carbohydrate supplier, it generates just as much carbon dioxide as it previously extracted from the air. The natural cycles are therefore CO_2-neutral and do not lead to an increase in concentration. However, this does not apply to the holiday trip to Italy and the transport of Italian spaghetti to other countries.

2.2.2 The Prime Suspect: Carbon Dioxide

In 1896, the Swedish scientist and Nobel Prize winner, Svante Arrhenius already worked out that doubling the carbon dioxide content (CO_2) in the atmosphere would increase the temperature by 4–6 °C [Arr96]. The connection between observed climate warming and the increase in carbon dioxide following industrialization was already being discussed in the 1930s, but at the time there was no way of confirming this connection.

It was not until the late 1950s that scientists were able to prove that the concentration of carbon dioxide in the atmosphere was increasing [Rah04]. Today there is widespread proof that the increase in carbon dioxide concentration is the main cause of the observed warming.

Table 2.1 The 10 countries on Earth with the highest energy-related carbon dioxide emissions (2015)

Country	Mil. t CO_2	Mil. inhab.	t CO_2/inhab.	Country	Mil. t CO_2	Mil. inhab.	t CO_2/inhab.
1. China	9041	1371	6.59	6. Germany	730	82	8.93
2. USA	4998	322	15.53	7. South Korea	586	51	11.58
3. India	2066	1311	1.58	8. Iran	552	79	6.98
4. Russia	1469	144	10.19	9. Canada	549	36	15.32
5. Japan	1142	127	8.99	10. Saudi Arabia	532	32	16.85
World	**32 294**	**7334**	**4.40**	133. DR Congo	3	77	0.04

Data: IEA [IEA17].

Fossil energy use is the main cause of the increase in carbon dioxide concentration. If we burn fossil energy sources, this is regarded as oxidation from a chemical point of view. This reaction releases heat. We therefore use the effect of heat that results when the carbon from oil, natural gas, and coal is combined with the oxygen in the air. The waste product we get in return is carbon dioxide – and we get it in enormous quantities: currently well over 30 billion tons every year. The average human produces about 4500 kg yr^{-1}. A corresponding amount of carbon dioxide would fill a cube with sides 13 m long, or around 2.3 million one-litre bottles.

As with energy consumption, the emissions in individual countries can vary a great deal (Table 2.1). For example, whereas someone from the Democratic Republic of Congo would tip the scales at only just 40 kg, i.e. 0.04 of a ton of CO_2, per year, in China it is almost 7 tons per head. In Germany it is more than 9 tons, and in the USA, more than 16 tons. If the carbon dioxide generated by Germans each year were dispersed over the country's entire land area, every German would be standing in 1 m of CO_2. In contrast, the carbon dioxide in the Democratic Republic of Congo dispersed over the country would not even form a layer 1 mm thick.

Yet, it is not that long since we have been able to claim with absolute certainty that the amount of carbon dioxide in the atmosphere is increasing on a yearly basis. The Mauna Loa Observatory in Hawaii has only been measuring the concentrations of carbon dioxide continuously since 1958. At that time the concentration amounted to 315.2 ppm, the following year it was already up to 315.8 ppm. The unit ppm stands for 'parts per million'. Thus, there were 315 parts of carbon dioxide per million parts of air. The small increase during the first year could also have been caused by a measurement error or natural fluctuations. It was not clear until the values began rising constantly in subsequent years that the amount of carbon dioxide was increasing – and even accelerating. By 2017 the CO_2 concentration had already risen to 407 ppm.

But, in comparison with the enormous size of the atmosphere, the high carbon dioxide emissions that occur when fossil fuels are burnt are minute. Furthermore, some of the carbon dioxide is absorbed again by the oceans and the plants. So, the question is how much can our emissions really change the composition of the atmosphere?

If we were to produce nitrogen instead of carbon dioxide when we use fossil energy sources, this would not pose a big problem. The reason is that our air consists of around 78% nitrogen and 21% oxygen but only 1% other gases – of which carbon dioxide makes up only a minuscule part. During the course of the Earth's history, the composition of the air has by no means been consistent. However, in the past few thousand years a balance of less than 300 ppm CO_2 has established itself. The proportion of carbon dioxide in the atmosphere was thus less than 0.03%. However, this is also the reason why we are even able to bring about any relevant changes as these small quantities can be increased comparatively easily.

A study of the climate of the past few millennia requires a different approach. In the regions with permanent ice, fresh snow falls on the ice surfaces every year. Large quantities of air are trapped between the ice crystals. The new snow masses that form each year alongside the old ones increase the pressure on the old snow and ultimately press it down into pure ice. In the process the air does not escape totally but instead remains as small bubbles trapped in the ice. Today these air bubbles can be examined using modern techniques of analysis. The deposits of snow and the creation of ice repeat themselves each year with a regularity welcomed by scientists. All it takes is to drill a hole into the ice and draw up ice samples from deep down. This provides a silent witness to the past. The deeper that one probes into the ice, the further back in history one is able to look.

Different drill core studies unanimously find that the concentration of carbon dioxide before the period of industrialization was around 280 ppm (Figure 2.8). Studies have shown that the proportion of carbon dioxide in the atmosphere is higher today than at any other point in the past 650 000 years (Intergovernmental Panel on Climate Change [IPCC], 2007).

Once it was finally proven that carbon dioxide emissions are increasing, climate models were developed to establish a correlation between the burning of fossil fuels and the increase in CO_2. Sources other than the emissions caused by humans were not considered as reasons for the increase. The models show that the CO_2 concentration could still more than double, depending on the extent of the future consumption of coal, oil, and natural gas.

This discovery raised another question, namely what the consequences of such drastic changes would be. In high concentrations carbon dioxide is unhealthy for humans,

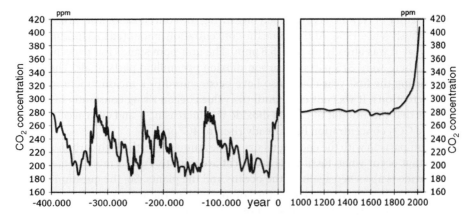

Figure 2.8 Time series of carbon dioxide concentration in the atmosphere over the last 400 000 years and in the recent past. Data: CDIAC, http://cdiac.ornl.gov.

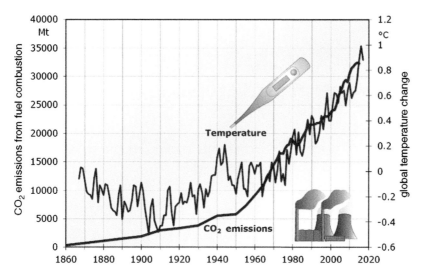

Figure 2.9 Progression of energy-related CO_2 emissions and global changes in temperatures since 1860 compared to the average between 1951 and 1980. Data: NASA, IEA.

and can even be life-threatening. However, to reach a level that would damage the health of humans, the concentration would have to increase a hundredfold and to cause any danger of suffocation, it would have to increase by a factor of 300. There is no way that we could reach such levels, even if we were to burn all the oil, natural gas, and coal available. At least any acute danger to the lives of humans can be ruled out.

Direct measurements of temperature have existed for well over 100 years. A comparison of global temperature changes with energy-related CO_2 emissions shows a significant connection between the two (Figure 2.9). Yet sceptics find fault with this argument and point out that the temperatures actually fell slightly between the 1940s and the 1980s. Today there is an explanation for this. Aerosols from high dust and soot emissions in the burning of fossil energy sources reduce the solar radiation on Earth. They act, so to speak, like sunglasses and thus create a cooling effect. Today modern filtering techniques have eliminated a major proportion of these emissions. This again decreases the sunglasses effect. However, CO_2 and the associated rise in temperature have remained.

As wood absorbs carbon dioxide during its growth, its burning then makes it carbon dioxide-neutral. However, this is only the case if the same quantity of plants is used and burnt as can grow back again. Tropical rainforests are currently being burnt at a much faster rate than their rate of regeneration. An area of woodland the size of the United Kingdom is disappearing every two years. Some of the forests being burnt are so large that they can easily be detected from space. If this does not change during the next few years, almost all the woodland in the world will be gone in about 100 years. During this time large quantities of carbon dioxide will be released, causing 10% of the greenhouse effect. However, the remaining CO_2 originates largely from the burning of fossil energy sources.

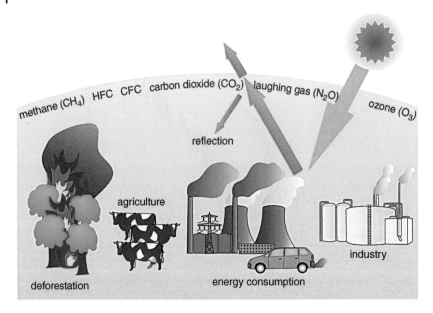

Figure 2.10 Causes of the anthropogenic greenhouse effect caused by humans.

2.2.3 Other Culprits

Fossil energy use is not the only culprit. Agriculture, the clearing of rainforests, and industry are also responsible for the additional greenhouse effect caused by humans (Figure 2.10).

Aside from carbon dioxide, there are other substances released through the actions of humans that cause the greenhouse effect. Methane, hydrofluorocarbons (HFCs), ozone, and nitrous oxide are some of the key ones. Although the concentration of these elements in the atmosphere is considerably lower than that of carbon dioxide, they have a much higher specific greenhouse potential.

The greenhouse potential of methane is 21 times that of carbon dioxide. This means that 1 kg of methane causes just as much damage as 21 kg of carbon dioxide. Methane is the second most important greenhouse gas. The concentration of methane in the atmosphere has already more than doubled because of human activity. Methane is also emitted during the exploitation of fossil energies. It escapes during natural gas extraction and coal mining, as well as from defective natural gas lines, and refuse disposal sites are another source (Table 2.2).

Interestingly, agriculture releases the largest proportion of methane. Livestock rearing, decaying biomass, and rice cultivation are the farming activities that produce the most methane. Unlike carbon dioxide, it is not integrated into a biological cycle. When ruminants digest green fodder, they create methane, which they belch into the atmosphere. A single cow or a herd would not pose a problem for the climate, but the number of cattle worldwide is now so large that they represent a real threat to the environment. Once again, it is the quantity that creates problems. Meat consumption per head has increased dramatically during the past few decades, as has the number of people, so that there are now more than 1.5 billion cattle on Earth.

Table 2.2   Characteristics of the most important greenhouse gases

	Carbon dioxide	Methane	HFC	Laughing gas (nitrous oxide)
Chemical notation	CO_2	CH_4	Various	N_2O
Concentration in the atmosphere in ppm[c]	407	1.85	<0.01	0.33
Concentration in the year 1750	280	0.75	0	0.270
Annual increase in concentration[c]	+0.7%	+0.5%	Varies	+0.3%
Global warming potential compared to CO_2[a]	1	21	>>1000	310
Persistence in the atmosphere in years	5–200	12	Varies	114
Reflection in $W\,m^{-2}$[b]	1.68	0.97	0.18	0.17
Percentage of anthropogenic greenhouse effect[b]	74%	16%	2%	6%

[a] Time horizon 100 years.
[b] 2011.
[c] 2017.
Source: Data: IPCC, NOAA, CDIAC.

Other greenhouse gases such as nitrous oxide (laughing gas) are created through farming due to an excessive use of nitrogenous fertilizers. Overall, agriculture contributes around 15% of the greenhouse effect caused by man.

Other key greenhouse gases include ozone, fluoride hydrocarbons (HFC and halons), water vapour, and sulfur hexafluoride (SF_6). Air pollutants increase levels of surface ozone. These pollutants originate in part from motorized traffic and, therefore, are also a result of the burning of fossil fuels. The ozone problem has become a familiar one because of the health risk associated with summer smog. However, what is less known is the impact of the ozone on the greenhouse effect. SF_6 is used in the electricity sector but the amount causing any greenhouse effect is relatively small.

Will the Hole in the Ozone Layer Keep Making Earth Warmer?

The ozone hole and the greenhouse effect are both global environmental threats. Yet they have very little relationship to one another.

With the ozone, a distinction must be made between the ozone layer at high altitudes and ozone at low level (at the surface of the Earth). The natural ozone layer with the 'good ozone' is located in the higher regions of the Earth's atmosphere – more precisely, in the stratosphere at an altitude of 15–50 km. The UV light of the sun is continuously converting air oxygen (O_2) into ozone (O_3) at this level. The ozone layer absorbs most of the dangerous UV radiation. Certain gases, such as chlorofluorocarbons (CFC) – from old refrigerators, air conditioners, and aerosol cans – decompose the ozone in the stratosphere. This is the reason why the ozone content in the ozone layer has decreased rapidly in recent decades. A hole in the ozone has formed, mainly over Antarctica. An increased amount of

UV light now reaches the Earth and is contributing to an increase in skin cancer. International agreements have limited the use of CFC. Its use has been forbidden in new plants in Germany since 1995. In the meantime, a slow recovery has been observed in the ozone layer.

Yet the ozone in the ozone layer is itself a cause of the natural greenhouse effect. For that reason, the destruction of the ozone layer has actually produced a slight cooling effect.

'Bad ozone' close to Earth, on the other hand, occurs due to the reaction of nitrogen oxides, oxygen, and sunlight and causes the notorious summer smog. Ozone acts as an irritant and in large concentrations is very toxic to humans. In the natural ozone layer above an altitude of 15 km this is not a problem, but it is a problem in the troposphere close to Earth. Furthermore, the tropospheric ozone close to Earth also contributes to the greenhouse effect. Unlike what happens in the ozone layer, the ozone concentration close to Earth is increasing. Therefore, it is the ozone close to Earth that is warming up the Earth.

CFCs were used as coolants in air-conditioning systems and refrigerators. This not only contributed to the greenhouse effect, but also damaged the ozone layer. As a result, the Montreal Protocol stipulated an end to CFC production worldwide, allowing long periods of transition. HFCs are often used today as a CFC substitute. These substances no longer use chlorine as an ingredient. Although this means that they can no longer damage the ozone layer, HFCs are still a problem for the greenhouse effect. The HFC cooling agent, R404A has a specific greenhouse potential of 3260. Therefore, 1 kg of R404A is just as harmful as 3.26 tons of carbon dioxide. But there are substitute substances available that are not damaging to the climate. Unfortunately, these are currently only being used in a limited number of areas.

Scientists have now developed a relatively good understanding of the different effects of individual greenhouse gases and other factors on the Earth's warming. The greenhouse gases that result from human activities increase the absorption and retention of heat in the atmosphere, resulting in global warming. Figure 2.11 shows the contribution of the different gases. The rise in solar activity is also causing a slight increase in radiation. As described earlier, aerosols (e.g. dust and smoke) released through human activities and the resulting cloud formations to which they contribute provide a cooling effect. Even the ozone layer and changes in land use are producing a slight cooling. If one juxtaposes the cooling and the warming effects, it is the warming that predominates. An increase in greenhouse gases and a decrease in aerosols through further improvements to air quality could even cause a considerable increase in warming over the next few years.

2.3   Outlook and Recommendations – What Lies Ahead?

No one can predict how many greenhouse gases will be emitted in the future. However, climate models can indicate the effects of different emission levels. For this purpose, climate researchers have designed different models that describe the effects of changes in greenhouse gas concentrations. On the basis of these models, they have developed complex computer programmes. To check a computer model, one first tries

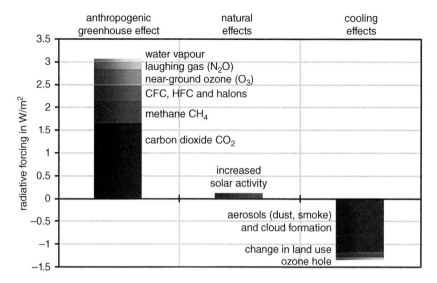

Figure 2.11 Causes of global warming. Data: IPCC [IPC07], www.ipcc.ch.

to reconstruct the changes of the past. Much progress has been made in this area of scientific endeavour, so that projections about future trends can now be made with a high degree of confidence.

The IPCC regularly publishes reports on climate change and studies the effects of climate. The United Nations Environment Programme (UNEP) and the World Organization for Meteorology (WMO) initiated the IPCC in 1988. The most competent climate researchers in the world are involved in the preparation of the IPCC reports and, as a result, this group has a good reputation internationally.

- www.pik-potsdam.de   Potsdam Institute for Climate Impact Research PIK
- www.ipcc.ch   Intergovernmental Panel on Climate Change
- www.unfccc.de   United Nations Framework Convention on Climate
                        Change (International specialist information on
                        climate change)

Climate sceptics regularly try to cast doubt on the results of climate researchers. Most of these are self-proclaimed climate experts with no relevant scientific background. They argue that climate researchers are developing and maintaining dubious theories in order to continue to obtain research funding. Climate sceptics are often in close contact with large energy companies that do business with coal, oil, and natural gas. If climate protection measures were to be effective, these companies would face severe economic disadvantages. The overwhelming majority of climate researchers are fully behind the scenarios produced by the IPCC.

It is virtually impossible to predict parameters such as population growth or the point when climate protection measures will be implemented. Therefore, the calculations in the models are restricted to certain scenarios. These indicate what can be expected in the best case and the worst-case scenarios. The conclusions of the climate researchers show that there is still time for us to have a drastic influence on future developments.

Climate change can no longer be stopped completely. However, decisive countermeasures could keep the consequences of climate change within manageable limits and largely save the Earth from vast perturbations to the climate. Of course, if there is no change in the coming years, climate change will very quickly reach a 'point of no return' and, in fact, would be unstoppable. Table 2.3 shows in which direction the course can still be set today. In the worst case, the entire Arctic ice melts. The melting of the Greenland ice sheet alone could cause sea levels to rise by 7 m in the next few centuries. This rise in sea levels will not be the same around the world. In Germany, the rise could be delayed. However, we will not be spared in the long term. Figure 2.12 shows the effects a sea-level rise of 7 m would have on Germany.

Table 2.3 Consequences of climate change – the climate can still be saved

This is how the climate could develop if it were protected worldwide	This is what threatens us in the worst-case scenario
• Global greenhouse gas emissions are reduced to zero by 2040.	• All fossil energy sources are almost used up and worldwide greenhouse gas emissions show a definite further increase.
• The average global temperature increases by more than 1.5 °C.	• The average global temperature increases by more than +8 °C.
• Sea levels rise significantly less than 1 m by 2100.	• Sea levels rise by 1–2 m by 2100.
• The Greenland ice sheet remains largely intact.	• The entire Greenland ice sheet melts away. In the long term, this alone will raise sea levels by 7 m.
• Even in the long term, sea-level rise remains in the range of a few metres.	• The Antarctic ice sheet also melts completely. In the long term the sea level rises by more than 60 m.
• Coastal regions and cities are saved through improvements in coastal protection measures.	• Entire coastal regions and cities such as Hamburg, Rotterdam, and Miami sink into the sea. Even cities on higher ground could be threatened in centuries to come.
• The Gulf Stream weakens only slightly.	• The Gulf Stream fails completely. The climate conditions in Europe undergo extreme changes.
• Heatwaves, periods of drought and extreme precipitation increase.	• Heatwaves, periods of drought and extreme precipitation increase dramatically and in some regions of the world threaten the foundations of life.
• Population migration due to climate change remains localized.	• The consequences of climate change trigger a major migration of people who flee their homes due to climatic conditions. Global tension rises dramatically as a result.

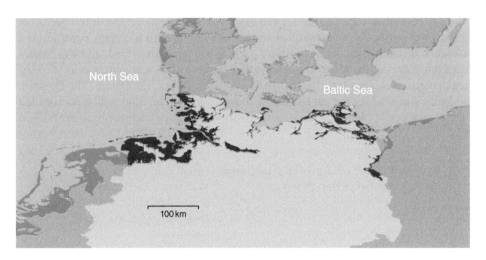

Figure 2.12 Threatened areas in Northern Germany if the sea level were to rise by 7 m in the long term. Graphic: Geuder, DLR.

2.3.1   Will it Be Bitterly Cold in Europe?

The American climate researcher, Wallace Broecker wrote an article in 1987 entitled 'Unpleasant Surprises in the Greenhouse'. The article caused quite a sensation at the time and warned about the potential for some seriously nasty surprises from the greenhouse effect: global warming could, paradoxically, turn Europe bitterly cold.

While studying the history of the climate from ice drill cores in the Greenland ice mass, climate researchers determined that the world's present climate is quite stable – but it has only been this way for about 10 000 years. The average annual temperature on Earth has never deviated more than one degree from the long-time average during this time. However, if one were to look back another 2000 years, one would find that major jumps occurred in temperatures even during short periods of only a few years. Until now it was always thought that temperature changes as well as the transition from ice ages to warming periods only happened very slowly. However, this opinion had to be corrected as a result of the drill core studies.

Today it is assumed that the climate tends to behave erratically on its own, thus making it a powerfully non-linear system [Rah99]. Linear systems are easy to understand. The stronger the attraction, the clearer the reaction. For example, a water tap approximates to a linear system. With two turns the amount of water that flows is double the amount from one turn. Climate is a complex non-linear system. It has a tendency to self-regulate and suddenly jumps into a different state when a critical point has been reached. The human body is an example of this self-regulation. It is able to keep its body temperature quite constant at 37 °C, even during major changes in ambient temperature. However, if a person becomes seriously hypothermic, their body temperature quickly drops at a certain point.

The Sahara is an example of very erratic climate change. Several thousand years ago the Sahara was green. The vegetation there absorbed moist air from the Atlantic and monsoon rains. Due to very minor changes in the Earth's orbit and in the inclination of the Earth's axis, the conditions for the monsoon rains deteriorated over a period of thousands of years. The vegetation managed to survive until a certain point around 5500 years ago. But then within a short period, the vegetation died, and the Sahara turned into a dry desert. A new stable state then developed, which continues until today. The populations of large areas had to flee due to the dryness of the terrain. The Egyptian Nile Valley was a refuge for the environmental refugees of the time, which contributed to the emergence of pharaonic high culture and was perhaps even the basis for biblical narratives.

www.scilogs.de/wblogs/blog/klimalounge	KlimaLounge Blog
www.pik-potsdam.de/%7Estefan	Website of the climate researcher Stefan Rahmstorf at the Potsdam Institute for Climate Impact Research

The origins of the erratic climate changes that occurred many years ago are suspected to lie in changes in ocean currents, which continue to have a major effect on our climate even today. For example, Berlin and London are located further north than Quebec in Canada. A comparison of temperatures shows that Western Europe is about 5°C warmer than it should be, considering its latitude. The reason for this is a gigantic self-regulating heat transport machine: the Gulf Stream.

In Central America the sun heats the water masses of the Caribbean Sea and the Atlantic. The Antilles Current and the currents from the Caribbean merge together to become the Gulf Stream, which is 50 km wide and transports enormous masses of warm saltwater northwards along the American coast. These currents are so large they can be seen from space. From North America the current then moves straight across the Atlantic, continues along the European coast towards Norway and ends in the European part of the North Sea. Some of the heat is absorbed by the air beforehand, and this air then moves with westerly winds towards Europe. This explains why winds from the West promise relatively mild temperatures to some parts of Europe.

In the North Sea between Norway, Iceland, and Greenland the warm saltwater merges with colder, less salty water, and cools down quickly. This water increases in density, becomes heavier, and consequently sinks to the bottom quickly. In the North Sea, every second 17 million cubic metres of water sink to a depth of 3000–4000 m. This corresponds to around 20 times the entire capacity of all rivers on Earth. The water masses then return to Central America as cold deep-water flow, to be heated up again there.

Due to climate warming, large quantities of meltwater (melted snow and ice) pour into the North Sea. This fresh water dilutes the warm saltwater. It has already been determined that the salt content has fallen considerably. If the salt content decreases even further, the weight of the water will not be sufficient in the future to allow it to sink down. The enormous Gulf Stream water pump would get out of step and could very suddenly come to a halt.

Climate researchers suspect this was precisely the cause of the temperature jumps 11 000–12 000 years ago. After the ice age enormous amounts of meltwater from the melting glaciers poured through the St Lawrence River into the North Atlantic. This diluted the sea water with fresh water, so that the uppermost layer of sea water became

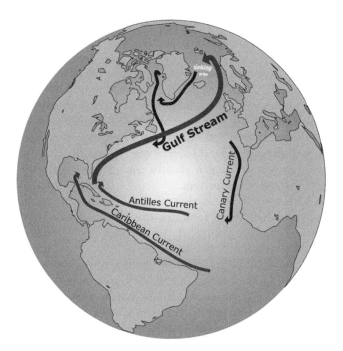

Figure 2.13 Principle of the Gulf Stream.

noticeably less salty and the water lighter. Therefore, the water no longer sank to the depths, despite strong cooling. The Gulf Stream was 'turned off' and for many years, large areas of Northern Europe and Canada were exposed to extreme cold and covered in ice. As a result, the amount of meltwater decreased again, the salt content of the water rose and the warm water heater, the Gulf Stream, started working again (Figure 2.13).

This is exactly the fate that we are facing again today. No scientist can accurately predict just how much global warming will have to increase before the Gulf Stream stops working. But the assumption is that the critical point will have been reached once worldwide warming increases by more than 3 °C. We already have increased the temperature by around 1 °C. The consequences for us would be catastrophic. Roland Emmerich showcased the topic in the blockbuster, *The Day After Tomorrow*. Even though many details are scientifically incorrect or strongly exaggerated, it nevertheless deals quite impressively with humanity's dangerous climate gamble. In reality, our climate would go crazy for years. Ultimately, it would be so cold and dry in Europe that any kind of farming would be impossible. Yet, according to the calculations of the climate researchers, this new climate state would then remain stable again for thousands of years.

2.3.2 Recommendations for Effective Climate Protection

Climate researchers consider a warming of 1.5 °C to be a critical threshold above which extreme and unmanageable consequences of climate change can occur. Only a very rapid reduction in greenhouse gas emissions can help to keep climate impacts within reasonable limits and avoid nasty surprises such as the disappearance of the Gulf Stream.

Of course, it is not possible to predict accurately to the year how much time will remain before the threshold is exceeded. In order to make statements that are as reliable as possible, climate researchers use sophisticated models on powerful mainframes to model the evolution of climate on Earth. The models have been continuously improved and can now also model many complex interrelationships. An example of this is ice albedo feedback. First of all, global warming causes the sea ice masses to defrost. Under the light-coloured ice, dark seawater appears, which absorbs sunlight better and the water heats up faster, resulting in even more ice melting.

However, despite all the improvements to the models, there are still some uncertainties. Therefore, in order to reproduce these correctly, the scientists work with probabilities where the greenhouse gas thresholds can be constrained by a certain date.

Climate sceptics like to use this to question the informative value of the models in general. However, there appears to be a lack of basic scientific understanding of future climatic changes. For example, if we try to determine when in spring the snow will have thawed in the low mountain ranges, we cannot predict the exact day when it will have completely disappeared. There is only a certain probability that it will have defrosted by mid-March, for example. However, this does not change the fact that the snow melts in spring and will have disappeared completely by a certain time with a certain time tolerance. Just because the researchers cannot give an exact date for the complete defrosting of the snow does not mean that the snow in the low mountain ranges does not thaw at all in summer.

As already described, carbon dioxide is by far the most important greenhouse gas. Between 1870 and 2011, around 1900 gigatons of carbon dioxide, or 1900 billion tons of carbon dioxide (Gt CO_2), were emitted worldwide.

Table 2.4 shows which carbon dioxide budgets are still available until certain temperature thresholds are exceeded. If we want to keep global warming below the critical value of 1.5 °C with a probability of 50%, we can only emit 310 Gt CO_2 from 2017. That is just one-sixth of all emissions emitted so far. But even with a 50% probability we risk a higher temperature rise than 1.5 °C.

Almost 38 Gt CO_2 were emitted in 2011. According to official figures, about 32 Gt of this was caused by the combustion of fossil fuels and industrial processes and about 5.5 Gt CO_2 by forestry and other land use. The slash-and-burn of the rainforest accounts for a very large proportion of this. In the meantime, total annual carbon dioxide emissions have risen to around 40 Gt CO_2.

If emissions remain constant at 40 Gt CO_2, the budget to keep the global temperature below 1.5 °C with a probability of 50% would be used up in eight years. If we start reducing emissions immediately, we should achieve zero emissions within 16 years.

If we accept that the temperature will rise above 1.5 °C but want to keep it below the much more critical threshold of 2 °C with a probability of 50%, then we would have 53 years left to reduce emissions linearly to zero (dashed curve in Figure 2.14). However, if emissions increase for a few more years, as in the past, or remain constant, the remaining time is noticeably reduced. Figure 2.14 shows this in terms of energy-related carbon

Table 2.4   Cumulative CO_2 emissions and remaining times to limit global temperature rise with different probabilities

Maximum warming	<1.5 °C	<2 °C	<3 °C
Probability of meeting the target	High|medium|low 66%|50%|33%	66%|50%|33%	66%|50%|33%
Cumulative CO_2 emissions since 2011 in Gt CO_2	400|550|850	1100|1300|1500	2400|2800|3250
Cumulative CO_2 emissions since 2017 in Gt CO_2	160|310|610	860|1060|1260	2160|2560|3010
Remaining time in years with constant emissions[a]	4|8|15	22|27|32	54|64|75
Remaining time in years with linear reduction in emissions[b]	8|16|30	43|53|63	108|128|151

[a]Constant CO_2 emissions of 40 Gt CO_2/a.
[b]Immediate stop of the increase and linear reduction of CO_2 emissions to zero.
Source: Data based on: IPCC [IPC15].

dioxide emissions. Emissions from slash-and-burn deforestation would also have to fall accordingly.

An immediate significant reduction in global emissions seems unlikely at present. The industrialized countries are currently causing the biggest per-head output of greenhouse gases. The developing and emerging countries justifiably want to catch up in terms of prosperity and energy use. That is why the industrialized countries must make the biggest contribution to reducing greenhouse gas emissions and serve as a model for

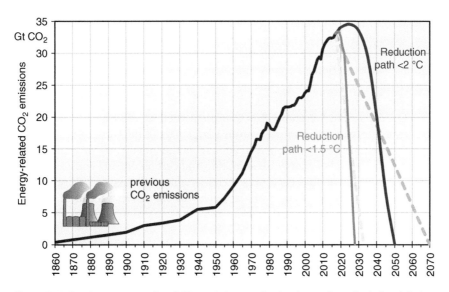

Figure 2.14   Previous energy-related CO_2 emissions and reduction paths to limit the global temperature increase to 1.5 °C or 2 °C with 50% probability.

the rest of the world. A responsible climate protection policy should therefore work towards reducing emissions in your own country to zero by 2040 at the latest. This probably makes it possible to limit the global temperature increase to below 2 °C. In order to comply with the 1.5 °C limit, emissions would have to fall to zero well before 2040.

To combat the greenhouse effect, radical technological change is needed worldwide. In the long term, the countries that will benefit most in economic terms are those that are leading the way in combating the greenhouse effect and are therefore the first to use the necessary sustainable technologies in their own countries.

2.4   A Difficult Birth – Politics and Climate Change

2.4.1   German Climate Policy

The realization that the consequences of an increase in global warming represents a previously unknown type of threat to mankind finally struck home for the politicians in the 1980s. At the time, the public was developing a keen awareness of the environment. Dying forests, respiratory illnesses due to air pollution, and the risk of nuclear power were topics that had a big impact on people. It was no longer possible to ignore global environmental problems.

In 1987 the German Bundestag (Lower House of Parliament) appointed the cross-party Enquete commission to deal with the issue of protecting the Earth's atmosphere. The commission studied the problems relating to climate change and drew up clear recommendations for reducing greenhouse gases [Enq90]. According to the report, emissions should fall by 25% by 2005 compared with 1990 levels, by 50% by 2020 and by at least 80% by 2050.

The first target of a 25% reduction by the year 2005 was adopted by the conservative government of Helmut Kohl. After reunification the policy was extended from West Germany to the unified Germany. After the change in government in 1998, the 'Red-Green' government, a coalition between the Green Party and Social Democrats, also adopted the climate protection targets for the year 2050.

The extension of the targets to the whole of Germany after reunification soon proved to be a wise move. According to statistics, German CO_2 emissions had fallen by as much as 17% by 2005. As a result, Germany likes to present itself as a forerunner in climate protection on the international stage. However, a large part of the CO_2 savings can be attributed to the effects of reunification. A major part of the industry in the former East Germany, with its high level of energy consumption and carbon dioxide emissions, collapsed. The output of greenhouse gases was thereby reduced to around half of what it had been, whereas in West Germany nothing much was changing in terms of climate protection. It was not until 2000 that the successes of the expansion of renewable energies had a noticeable impact on emissions.

However, the measures introduced were not adequate for an effective climate protection policy that would make the targets realistic. The first target of a 25% reduction by 2005 was missed by a long shot (Figure 2.15). The reduction target for 2020 was lowered to 40% and for 2030 to 55%, yet, in view of the large global increase in carbon dioxide emissions in recent years, significantly larger reductions would be required.

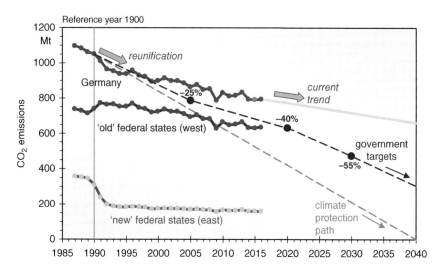

Figure 2.15   Energy and process-related carbon dioxide emissions in Germany.

However, the current measures are hardly sufficient to meet even the very conservative government targets for 2020 and 2030. The gap between government goals, an effective climate protection path, and real developments threatens to grow even further.

2.4.2   International Climate Policy

In 1979 the main theme at the first UN Conference on World Climate in Geneva was the threat to the Earth's atmosphere. However, many special-interest representatives had fundamental doubts about the threat. So, it was initially agreed that further research would be needed before any particular measures could be considered.

In Rio de Janeiro in 1992 the members concluded for the first time that measures to protect the environment were now absolutely essential. The first targets for reducing greenhouse gas emissions were then agreed in 1997 for industrialized countries in the Kyoto Protocol. However, these goals were anything but sufficient for effective climate protection. Global greenhouse gas emissions continued to increase despite the Kyoto Protocol. The Paris Climate Protection Agreement of 2015 aims to stabilize the global rise in temperature to as little as 1.5 °C, if possible, involving all the countries of the world. To achieve this, however, greenhouse gas emissions would have to be reduced to zero well before the middle of the century, as already explained.

The Paris Climate Protection Agreement

The first world summit on climate, the UN Conference for Environment and Development (UNCED), took place in Rio de Janeiro in 1992. One result of the conference was the UN Climate Framework Convention on Climate Change (UNFCCC), where the signatories made a commitment towards preventing dangerous anthropogenic disruption to the climate system, slowing down global warming and lessening its consequences.

This rather vague agreement was followed by the Kyoto Protocol, which was drawn up in 1997 at the third conference of the treaty states of the UNFCCC in the Japanese city of Kyoto. It envisaged different reductions for the industrialized countries. The agreement expired in 2012 and only a few states reached their targets. Global greenhouse gas emissions continued to rise significantly despite the Kyoto Protocol, as the emerging and developing countries were not included in the agreement and many of the signatory states failed to honour their commitments.

In 2015, a new agreement was finally negotiated at the 21st Conference of the Parties to the UN Framework Convention on Climate Change (COP21) in Paris. The agreement came into force on 4 November 2016. The main goal is to keep the global temperature increase well below 2 °C and preferably not to exceed 1.5 °C. The agreement has been ratified by all states. However, under President Donald Trump, the US has announced that it will be the only country to withdraw from the agreement in 2020. Within the framework of the agreement, each country must implement self-defined measures that work towards the common goal and report every five years and, if necessary, intensify the measures. However, all the measures adopted so far at national level are still insufficient to keep global warming below the agreed limit values. Some countries, such as Germany, for example, are also clearly missing their self-imposed targets. It is to be hoped that the continuous negotiation process and the reporting obligations will increase the pressure continuously to achieve the goals and to continuously tighten the requirements in the sense of the agreement.

In the many industrialized countries, so far, the fight for our climate has been very tentative. Countries that have recorded definite reductions in greenhouse gas emissions include those that were formerly in the Warsaw Pact. The reason for this is due more to the economic upheavals of the 1990s than to a consistent policy on climate protection. (Figure 2.16). Despite the positive effects on emissions from reunification, Germany is only in the middle of the field in terms of emission reductions. Countries such as Canada and Australia, on the other hand, have even increased their emissions significantly.

Even though a number of Western countries such as Denmark, Sweden, Great Britain, and Finland have achieved significant savings, this is not enough for effective climate protection. The backlog in implementing alternative technologies in developing and emerging countries, many of which have more than doubled their emissions since 1990, is currently eating up all the savings we had previously made. Developed and developing countries can therefore only achieve effective climate protection with joint efforts. With a little hope, one can look to China, that has recently begun to make great efforts to protect the climate effectively.

However, the varying degrees of change in the individual countries in just a few years show that much is possible in terms of climate protection. With really serious climate protection efforts, it should still be possible to achieve the necessary reductions in greenhouse gases to limit global warming to well below 2 °C.

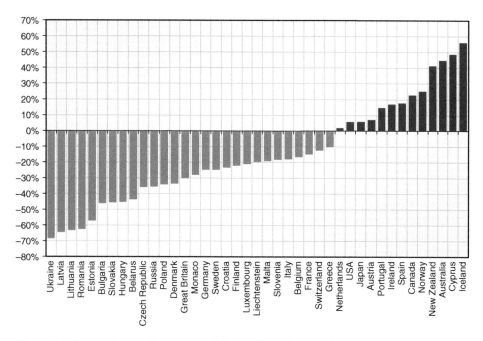

Figure 2.16 Changes in greenhouse gas emissions with no change in land use between 1990 and 2015. Data: UNFCCC [UNF17].

2.5   Self-Help Climate Protection

The governments of most countries today are taking a half-hearted approach to addressing the issue of climate protection. Certain special-interest groups would invariably suffer financially if consistent policies were pursued in this area. Therefore, these groups are trying to use their political influence to delay any serious implementation of climate protection measures. Reference is also often made to other nations that do even less for climate protection.

In extreme life situations such an approach seems unthinkable. Imagine a group of people sitting in a boat with a severe leak. Only by working together could the leak be sealed, and a disaster averted. But no-one does anything and everyone is discussing that one would only move towards the leak once everyone else gets up. Since this does not happen, everyone watches together as the boat slowly sinks. It is very likely that if one of them rolls up their sleeves and starts to plug the leak, the others will step in, one after the other, to help to prevent the catastrophe.

Therefore, climate change is no reason to lapse into lethargy. On the contrary: everyone has the chance to make a contribution towards protecting the climate. Savings of 50%, 80%, or even 100% in greenhouse gas transmissions are easily achievable right now. More and more people will then follow the good example, and the results of their efforts will help to protect the climate. But a single measure on its own is not enough to stop global warming. Instead it will take a combination of many different kinds of measures. These measures also drive investment in sustainable technologies and thus secure our long-term prosperity. Renewable energies in particular can play a major role here, as this book will show.

Active climate protection does not demand any major sacrifices. If everyone tried to implement the following measures, it would be enough to save the climate:

- Avoid buying products made of tropical woods that do not come from sustainable sources. The Forest Stewardship Council (FSC) label can help in a purchase decision.
- Avoid equipment and appliances (refrigerators, air conditioning units, heat pumps) containing coolants with HFCs because they contribute to the greenhouse effect. If such equipment is not available without HFCs (e.g. car air-conditioning systems or heat pumps), the buyer should put pressure on the manufacturer.
- Be consistent in recycling, reduce meat consumption, and increase the use of organic and local products.
- Avoid unnecessary consumption and buy durable equipment and products.
- Use public transport, bicycles, long-distance coaches, or trains, buy electric cars and avoid air travel.
- Be consistent in saving energy in all areas (see Chapter 3).
- Switch to an independent green energy supplier.
- Build and finance new renewable energy systems and installations (see Chapters 5–14).

3

From Wasting Energy to Saving Energy and Reducing Carbon Dioxide

In modern industrial societies energy is still available at relatively moderate prices and without any restrictions. Whenever we want it, we can consume as much energy as we like. However, the general increase in energy prices in recent years has brought about a new awareness of the value of energy. High energy consumption is making more and more of an impact on our finances. Yet the equation is very simple: saving energy also means saving money. But the concept of saving energy is not totally new.

Until the 1970s it was accepted that real economic growth and increasing prosperity would also require the use of more energy. The concept of saving energy developed when the oil crisis of the 1970s led to an explosion in oil prices and put a brake on growth. Numerous tips, appeals, and stickers in the 1980s were aimed at encouraging citizens to save energy, and the trend towards stopping energy waste was indeed successful at the time. As prices started dropping again in the 1990s, the original goal was largely ignored, and energy was again thoughtlessly wasted (see Figure 3.1).

3.1 Inefficiency

High energy consumption is not a necessary prerequisite for maintaining our prosperity and our standard of living. In fact, energy use is coupled with enormous losses. Around 35% of the primary energy used is already lost in the energy sector through power plant waste heat or during transport even before it reaches the consumer. Various devices and machines then produce the desired usefulness, such as light, heat, and driving power for machines and vehicles. This too leads to high losses. Light bulbs and vehicle combustion engines are particularly inefficient and transform between 80% and more than 90% of the energy used into undesirable waste heat. Even 'useful' energy is often wasted, for example when lights are left on in empty rooms, poorly insulated buildings are heated, or people drive round and round the block to find a parking space (Figure 3.2). If one looks at all the losses incurred, it emerges that, at best, only 20% of original primary energy is being used efficiently (Section 3.1).

When it comes to certain aspects of our lifestyles in industrialized societies, the percentage of energy used efficiently is even smaller still. However, the sort of person who tries to talk friends out of taking long flights or buying a new car with all the extras, or who keeps telling his family to turn down the central heating, tends not to be very popular. The choice of one's own lifestyle, finances permitting, is considered one of those

Figure 3.1 During the 1980s, energy saving was an important topic in Germany, and the "I'm an energy saver" sticker was a familiar sight. Rather ironically, it was particularly popular on gas-guzzling cars.

individual freedoms that no one else has a right to influence. Therefore, saving the environment should not be about looking for scapegoats, but instead about seeking solutions that meet both lifestyle and environmental needs.

This does not mean that everyone can abuse his or her right to a certain lifestyle and thoughtlessly treat energy and the environment with disrespect. People are acting irresponsibly if they knowingly refuse to accept a technology that would enable them to retain their lifestyle but at the same time considerably reduce energy consumption and save the environment. This also applies to policies that do not strive towards a speedy introduction of the most optimal technologies.

Many small steps that could help create an environmentally compatible society are also not being implemented due to a lack of knowledge or appropriate awareness. Many of the problems related to energy use are extremely complex. Optimal solutions to these problems often depend on a number of different factors.

For example, one issue that is often debated is whether a gas or an electric cooker is more efficient. Compared to an electric cooker, a gas cooker produces more waste heat. Anyone who has ever scorched a potholder on a gas hob will be able to confirm this. Yet, gas has the reputation of being the ecologically better alternative. Conventional coal-fired, natural gas and nuclear power plants for the generation of electricity do not work efficiently. More than 60% of the primary energy used in these plants is lost as waste heat (Figure 3.3). If the electricity used to boil a litre of water comes from a coal-fired power plant, it is releasing 156 g of carbon dioxide. In contrast, burning natural gas on a gas cooker releases about 56 g of carbon dioxide. However, the losses that occur when gas is transported from the place of extraction to the end consumer are also a problem. Natural gas essentially consists of methane, which is considerably more damaging to

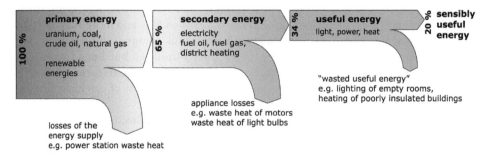

Figure 3.2 In Europe, around 80% of energy is lost or not used efficiently during transport or conversion.

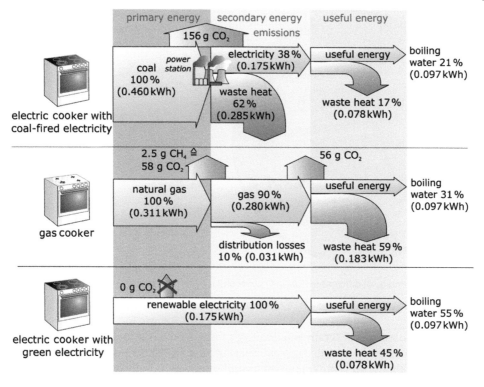

Figure 3.3 Energy and environmental balance of boiling water using an electric versus gas cooker.

the environment than carbon dioxide. Therefore, even the loss of a few grammes can be harmful. If transport losses are at the rate of about 10%, a gas cooker is still causing a lower rate of greenhouse gas emissions than an electric cooker using electricity generated by coal. However, the pipelines in the areas where natural gas is exploited are sometimes in very poor condition. The use of natural gas from these sources can completely cancel out the advantages of gas cookers.

In Germany, a gas cooker is a better alternative than an electric one, if the electricity comes from a conventional energy supplier. 'Green' electricity suppliers offer carbon-free electricity from renewable energy plants. Using this kind of electricity makes an electric cooker more environmentally-friendly than a gas cooker. In Norway, all electricity is supplied by renewable energy plants. In this case, electric cookers are generally the first choice. Finally, the use of an electric kettle to boil water is an efficient way to save energy and reduce carbon dioxide and puts the user ahead of the game.

- www.energie-vision.de   Ok-Power label for 'green electricity'
- www.gruenerstromlabel.de   Green power label
- www.energieanbieterinformation.de   Information on changing supplier
- www.vergleich.org/echter-oekostrom   Suppliers of genuinely green electricity

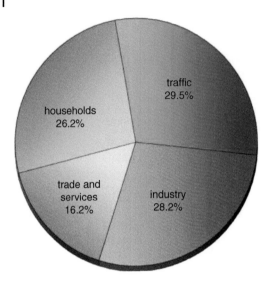

Figure 3.4 Percentage of different sectors in final/secondary energy consumption in Germany. Data: [BMWi18], 2016.

Anyone who worries generally about saving energy should first look at the sectors that use the most energy. What will come as a surprise to many is that private households in Germany use almost the same amount of secondary energy as industry and the transport sector do (see Figure 3.4). Over the last 20 years, no significant savings in final/secondary energy consumption have been achieved. In general, there are major fluctuations from year to year. But these are mainly due to the weather. Consumption is lower in mild winters and higher in cold winters.

One reason for the stagnating energy consumption of households is the increasing number of electrical appliances. This increase is particularly noticeable in the areas of communication and entertainment electronics. But there are also many hidden energy guzzlers that have unnecessarily high energy use in standby mode. As the possibilities for saving energy in the household and transportation are particularly easy to implement, these options top the list of any analysis on energy savings.

3.2 Personal Energy Needs – Savings at Home

3.2.1 Domestic Electricity – Money Wasted

Europeans take the supply of electric energy so much for granted that we find it hard to imagine doing without electricity even for a short time. Televisions, telephones, computers, lights, refrigerators, washing machines and even heating do not function without electricity. It is hard to grasp, then, that around a billion people, one-seventh of the world's population, have no access to electric power.

In Germany, an average three-person household consumes around 3900 kWh of electricity at a cost of about €1100 yr^{-1}. More than 10% of the carbon dioxide emissions from energy use in Germany come from the electricity needs of private households [UBA07].

Furthermore, inefficient electrical equipment accounts for a considerable percentage of electricity consumption. Anyone looking for ways to cut back on energy consumption can quickly find ways to save up to 30%, which equates to more than €300 a year. Therefore, protecting the climate can also pay off financially. An average US household consumes even more than $10\,000\,\mathrm{kWh\,yr^{-1}}$. Hence, the potential for saving is much higher there.

Standby Losses – Power Destruction Par Excellence

Many electrical devices work at low voltages. A transformer transforms down the mains voltage for this purpose. Most devices have a switch-off mechanism, but it usually only switches off part of the electronics in the lower voltage area. The transformer and major parts of the device electronics remain connected to the grid and continuously consume electricity, even when in a switched-off state; these are known as standby losses. As a result, over the course of a year a considerable amount of energy can be consumed even by low-power devices such as televisions. Power wasters like this can actually be identified using an energy consumption measuring device.

There are very few cases where, for technical reasons, a device should not be completely unplugged from the mains. Mains switches are a few cents more expensive than low-voltage switches. When large numbers of devices are involved, a manufacturer can easily save thousands of euros without fearing a loss of sales by installing the cheaper switches. Standby loss is currently not a major consideration in purchase decisions.

Computers and communication and entertainment electronics currently account for around one-fifth of electricity consumption in Germany and one-tenth in the USA. This kind of equipment often wastes electricity for no reason. Many devices have high standby consumption, which means they consume electricity even when turned off. A device with standby consumption of only 5 W uses a total of $43.8\,\mathrm{kW\,h}$ and costs €12 in electricity per year – just for the period when it is switched off and not being used at all. Some, usually older, devices even have standby losses of over 30 W. On average each household in Germany wastes around €120 $\mathrm{yr^{-1}}$ due to standby losses. The German federal environment office estimates that open-circuit losses in Germany cost about 5 billion euros per year and produce 14 million tonnes of carbon dioxide. This is more than seven times the amount of carbon dioxide emitted through energy use by the 20 million inhabitants of Mozambique.

This is something that is very easy to remedy. By using a switchable multipoint plug for their computer, television and stereo system, consumers can reduce their open-circuit losses in a switched-off state to zero. When buying new devices or equipment, consumers should ask questions about standby consumption values so that more pressure is placed on manufacturers to fit power-saving switches.

Around 10% of the electricity used in private households in Germany is for lighting. In the USA the percentage is about the same (see Figure 3.6). Here, too a great deal of energy, and money, is wasted. Due to inertia and unfounded prejudice against energy-saving bulbs, millions of conventional electric light bulbs are still being used. This is not only damaging the environment but also costing money. Modern low-energy or LED light bulbs produce the same amount of brightness using 80% less electricity. In contrast

Figure 3.5 Low-energy light bulbs save energy, carbon dioxide and cash. LED lamps are particularly environment-friendly substitutes for conventional incandescent lamps. Photos: OSRAM, Siemens press photo.

to first-generation low-energy light bulbs, advanced LED lamps do not contain mercury, and there is no irritating warm-up delay. An 11-W LED lamp or low-energy bulb saves about one cent an hour. If this kind of light bulb burns for two hours each day, the savings amount to seven euros a year. Over its lifetime a low-energy light bulb can save €150 and 300 kg of carbon dioxide. Therefore, the higher purchase cost is rapidly recouped. Good-quality LED lamps or low-energy light bulbs also tolerate frequent on- and off-switching well. Tiny, dimmable light bulbs open up possibilities to all kinds of uses (Figure 3.5). A consistent climate protection approach should include a gradual replacement of all light bulbs with LED lamps.

Electric water heating is relatively expensive. Anyone who has an electric boiler should switch it off when leaving the house for long periods.

A considerable amount of energy can also be saved with large electrical appliances, such as washing machines, dishwashers, refrigerators, and freezers. Efficiency should be an important criterion, especially in the selection of new appliances. Energy efficiency rating labels and the consumption values included with these labels provide an indication of the efficiency and economy of use of the different appliances.

Unfortunately, the efficiency ratings have not always provided clarity in the past. The original intention was to assign the letter A to efficient devices and less efficient devices the letters B to G. Since the devices have been increasingly improved in recent years, those having an efficiency rating A had become average at best. Better devices were assigned new efficiency ratings from A+ to A+++. It was not easy to keep track. The best washing machines were able to beat the values required for A+++, while for a long

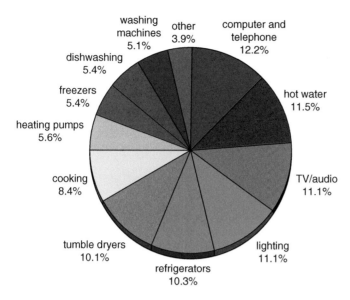

Figure 3.6 Breakdown of electricity consumption of private households. Data: [Ene06].

time A+ was the best for LED lamps. In order to end the confusion, the EU has decided to return to the old labelling with the letters A to G from 2019.

In any case, if you always buy the most efficient appliances, you can save a lot of money and carbon dioxide over their service life. If you are unsure about how much energy a device or appliance consumes, there are energy-saving monitors available for hire or purchase, which test appliances for energy waste. It is advisable to take particularly inefficient appliances out of service as quickly as possible.

- www.spargeraete.de   List of efficient households appliances
- www.ecotopten.de   Overview of efficient appliances
- www.die-stromsparinitiative.de   Electricity saving tips

A savings potential also exists in the way household goods are used. If a refrigerator is standing right next to a cooker, it must work harder because of the heat loss from the cooker. Refrigerators with poor ventilation and iced-up freezer compartments also use a large amount of extra energy. A lid on a pot when cooking can save an appreciable amount of energy, as can a pressure cooker. Selecting the lowest temperature possible on a washing machine or a dishwasher saves energy and money. With tumble dryers the laundry should first be spun at the highest spin speed available. But the good old clothes line is the best option for saving energy.

The following energy-saving tips summarize the key points discussed:

- Track energy guzzlers using an energy consumption measuring device.
- Switch off unnecessary electrical devices and lights.
- Switch off all devices with standby mode using a switchable multipoint switch.

- Replace incandescent light bulbs and halogen lamps with LED lamps.
- When buying electrical goods, pay attention to their energy consumption ratings.
- Always buy the most efficient domestic appliances.
- Do not place freezers next to heat sources (ovens, radiators).
- Try to thaw frozen goods in a refrigerator.
- Defrost refrigerators and freezers regularly.
- Only run washing machines when they are full and operate them at the lowest possible temperature and set to a high spin speed.
- When cooking, use lids on pots and frying pans, or use pressure cookers.

3.2.2   Heat – Surviving the Winter with Almost No Heating

In many countries the lion's share of secondary energy consumption in private house-holds comes from heating. Around three-quarters of secondary energy in households is used for this purpose. However, reducing heat in a home does not necessarily mean hav-ing to cope with frosty temperatures. With good insulation and modern building tech-nology, pleasant room temperatures can be achieved with energy savings of up to 90%. In other words, 10 energy-efficient buildings can be kept warm with the same amount of energy it takes to heat the average poorly insulated old house. At the same time the carbon dioxide emissions and heating costs drop to one-tenth.

Many people live in rented accommodation. This can cause a dilemma, because energy-saving measures are usually linked to investment. This is a cost that the land-lord first has to bear, with the renters becoming the beneficiaries – and because they did not pay for the measures in the first place, they do not have any particular incentive to implement them. However, from the standpoint of saving energy, even actual homeowners lag behind, when it comes to making use of the options that are available.

Yet, not all energy-saving measures cost money. The following changes to heating behaviour can save a considerable amount of heat energy and, consequently, reduce heating costs:

- Do not select a room temperature that is higher than necessary. Each extra degree consumes around 6% more heat energy.
- Turn down the heat at night, and when no one is at home.
- Close blinds, shutters, and curtains at night.
- Avoid leaving windows open all the time to air rooms; instead briefly open windows completely a few times a day to bring in fresh air.
- Do not enclose, block, or hide radiators behind curtains.

Even minor investments can help reduce energy use considerably. Appropriate mea-sures can include draught-proofing windows and doors and upgrading thermostat valves.

In the case of three-litre and passive houses, heating requirements can be slashed to as little as one-tenth of what is normally needed in conventional houses. The heating required in an average apartment block in Central Europe is around 150–200 kilowatt hours per square metre of living area per year ($kWh/(m^2 \, a)$). Old buildings with poor insulation can even show usage values of over $300 \, kWh/(m^2 \, a)$. In contrast, a three-litre

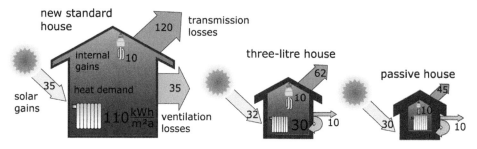

Figure 3.7 Comparison of heating energy demand and heat loss in houses with different insulation standards in kilowatt hours per square metre of living area per year in Central European climates (kWh/(m² a)).

house only needs $30\,\text{kWh}/(\text{m}^2\,\text{a})$, whereas a passive house gets by with even less than $15\,\text{kWh}/(\text{m}^2\,\text{a})$ (Figure 3.7).

Not All Low-Energy Houses Are Equal

Anyone interested in building a low-energy house will first have to grapple with a variety of different terms and concepts that sometimes confuse even the experts. The following should help to unravel some of this confusion.

…

Low-energy house. These are especially well insulated buildings that have low heating energy demand. Most countries have not legally specified this concept.

Three-litre house. A three-litre house colloquially is a house that requires around 3 l of heating oil per square metre of living area per year (approx. 30 kWh) in Central European climates. This corresponds to a primary energy requirement of around 60 kWh/(m² a), which also includes the efficiency of the heating system and the hot water system.

Passive house. A passive house can almost manage without any heating at all. The annual heating energy demand is less than 1.5 l of heating oil (approx. 15 kWh) per square metre of living area. This equates to a primary energy demand of less than 40 kWh/(m² a).

Building insulation and window type have an important effect on heating energy demand (Figure 3.8). The U-value is a comparison value for the quality of heat insulation. This value indicates the heat loss per square metre of wall and window space and per degree of temperature difference between indoors and outdoors. Low U-values therefore mean low heat losses.

Compared to non-insulated outer walls, walls with high-value heat insulation can achieve a reduction of the U-value by more than a factor of 10 in an optimal case. Conventional insulation against heat loss of around 20 cm would be required. This is easily achievable with prefabricated houses using a wood frame construction. Vacuum-insulated materials with a thickness of around 2 cm that provide the same insulation results have recently been developed. With this type of insulation a core material of pyrogenic silicic acid is packed and evacuated into an airtight foil. It is important that

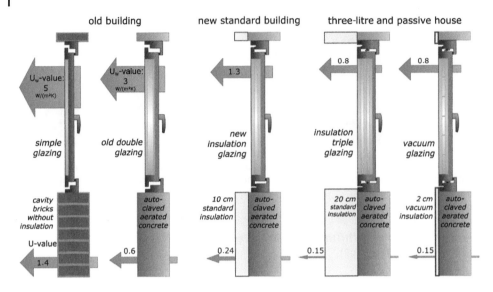

Figure 3.8 Effect of window type and insulation on heat loss.

the special foil of the vacuum heat insulation prevents any air from seeping in over long periods of time. However, this insulating material is quite expensive.

Along with the insulation of the walls, the structure of the windows is also an important element. Compared to conventional thermal double glazing, triple glazing can reduce heat loss from windows by over 40%. Vacuum glazing is a new technology on the market in which air is evacuated from the gap in the double glazing to create a vacuum. A perfectly sealed edge prevents any penetration of air. Small spacers set between the layers of glass provide the necessary stability, so that the exterior air pressure does not crush the two glass panels together. Spacers made of glass are hardly noticeable.

In addition to the type of glazing selected, the type of window frame is also important. Therefore, for the U-value a differentiation is made between a U_G-value (g for glass) for the actual glass panel and the U_W-value (w for window) for the whole window, including the frame. Some manufacturers only provide a general U-value. In most cases, they mean the U_G-value.

Windows are not just sources of heat loss. They also let in sunlight, which in the winter helps to heat rooms. For an optimization of solar energy gain, south-facing windows in colder climates in the Northern hemisphere should be as large as possible, whereas those facing north should be smaller. Exterior shutters or blinds on the sunny side are important to prevent well-insulated buildings from becoming too hot in the summer.

Buildings with optimal insulation are comparatively airtight. For good air quality in the winter, fresh air should be let into rooms frequently. But airing rooms also causes considerable heat loss. This is remedied through the use of controlled building ventilation systems (Figure 3.9). Fans blow fresh air into living areas and extract stale air from kitchens and bathrooms. The stream of fresh air is run from outside through a cross-heat exchanger past the exhaust air. The stale air emits up to 90% of its heat to the cold, fresh air from outside, thereby keeping the heat in the building.

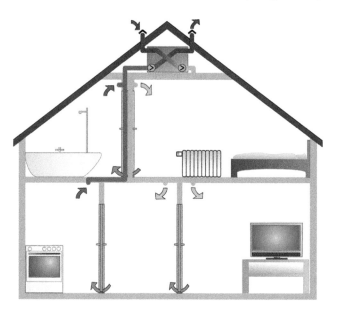

Figure 3.9 Principle of control ventilation with heat recovery.

Controlled ventilation is often perceived as having a negative effect on living condi-tions, but the opposite is actually the case. A gentle draught from such a system is barely noticeable. Constant optimal ventilation prevents dampness forming in walls or mould building up. An air filter in the ventilation system will keep out some of the pollutants in the outside air and also make it difficult for insects to get inside. Of course, anyone who feels that it is necessary can open the windows anyway – even if this is really no longer necessary from a practical point of view.

An air supply system can also be combined with a ground heat exchanger. This involves laying a pipe through the soil in the garden to supply incoming air. In winter the soil heats the fresh air and in summer it cools it.

An ultra-high energy-saving house requires an outlay of additional costs, but over the years these costs pay for themselves as energy prices rise. In some countries, such as Germany, banks offer particularly good low-interest loans to promote energy-saving measures in new houses and in the renovation of old buildings.

- www.energiesparhaus.at   Independent energy-saving advice
- www.passiv.de   Passive House Institute
- www.kfw-foerderbank.de   Information on KfW Bank programmes

In principle, those who want to keep heating requirements carbon-neutral have the following options:

- solar thermal systems;
- biomass heating;

- heat pumps using electricity from renewable energies; and
- heating systems based on renewable methane or hydrogen.

Solar thermal systems usually supplement other types of heating. Later chapters in this book discuss these different variants in detail. As the potential for certain possibilities such as biomass heating is limited in many countries, the other options for saving energy described above should be implemented as far as possible before renewable energies are used.

3.2.3   Transport – Getting Somewhere Using Less Energy

The transport sector is responsible for about one-fifth of carbon dioxide emissions from energy use. Whereas it seems that major savings can be achieved relatively quickly with electricity use and in heat generation, the transport sector is more problematic. In recent years increased mobility and the travel bug have managed to cancel out any savings in the fuel consumption of cars and aeroplanes. Cheap airlines that sometimes offer plane tickets for the price of an underground fare, and the trend towards petrol-guzzling SUVs, are prime examples of this development. Added to this are the even longer distances involved in transporting goods around the world as a result of globalization. Turning back the clock and reducing the number of cars, planes, trucks, and ships would be a hopeless undertaking.

Modern economies and modern lifestyles have become so fast-paced that they demand increased mobility. Holidays tend to be short, so air trips are necessary in order not to eat into travel time. Weekend breaks by air provide relaxation in hectic schedules. Our economic growth also is based on high levels of exports and reasonably priced raw materials, both of which are associated with long-distance transport.

Nevertheless, to reduce carbon dioxide emissions, we simply must transfer them from A to B using less energy. In the medium term, carbon-free transport must be available to cut out emissions completely.

Selecting the right means of transport is the best option for a quick reduction in energy requirements and emissions in traffic. For example, per passenger kilometre the secondary energy requirement of a train is less than one-fifth that of a car. To compare the two modes of transport, the energy consumption of the train is divided by the average number of passengers.

While the railways in Norway or Switzerland are already almost climate-neutral due to the high proportion of hydropower, rail traffic in Germany still generates significant emissions (see Figure 3.10). This is because over 30% of the electricity used by trains comes from climate-damaging coal-fired power plants. In 2016, around 17% of the electricity came from nuclear power plants and more than 40% from renewables.

The calculations shown here are based on average values. The average carbon dioxide emissions of new cars in Europe are currently around 120 g of CO_2 per kilometre. Older vehicles and thirsty new cars have higher emissions, whereas economical small cars sometimes have considerably less. In addition, greenhouse gas emissions are generated proportionately during oil extraction, fuel production and fuel transport.

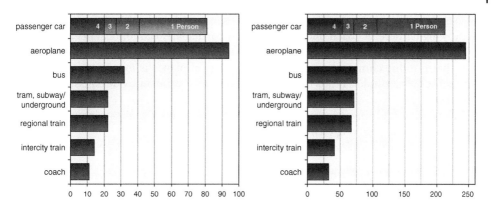

Figure 3.10 Energy consumption and greenhouse gas emissions per person for different means of transport with an average use load (data for trains is estimated from the German electricity grid).

Not only the type of car, but also the way in which it is driven has a major impact on how much petrol is consumed. The following energy-saving tips will enable savings of up to 30%:

- Check tyre pressure frequently and inflate to at least the pressure recommended by the car manufacturer for a fully loaded vehicle.
- Change gear as early as possible, drive at a uniform speed and always anticipate what is ahead of you.
- Restrict maximum speed on motorways.
- Do not transport unnecessary ballast or unneeded roof luggage racks.
- Switch off air conditioning and other energy-consuming devices when they are not needed.
- For short distances try to walk or go by bicycle.

Determining Carbon Dioxide Emissions Based on Fuel Consumption

Petrol $\boxed{}$ $\dfrac{\text{litres of petrol}}{100\ \text{km}}$ $\times$ 23.7 $\dfrac{\text{g } CO_2}{\text{litres of petrol/100}}$ $=$ $\boxed{}$ $\dfrac{\text{g } CO_2}{\text{km}}$

Diesel $\boxed{}$ $\dfrac{\text{litres of diesel}}{100\ \text{km}}$ $\times$ 26.5 $\dfrac{\text{g } CO_2}{\text{litres of diesel/100}}$ $=$ $\boxed{}$ $\dfrac{\text{g } CO_2}{\text{km}}$

The capacity utilization of a car has the biggest effect on energy use. On average three to four seats in a car are empty when it is being driven. With four people in a vehicle and careful driving, the carbon dioxide emitted per person can be even lower than with a train.

Capacity utilization also plays an important role with other forms of transport. Trains have particularly low energy consumption. As the utilization of local public transport is on average lower than that of long-distance transport, carbon dioxide emissions are also slightly higher here. A coach filled to capacity is particularly economical in energy use.

Air traffic produces the highest carbon dioxide emissions. Long-haul flights cause the most harm to the climate, because the exhaust gas of planes is more damaging when emitted at high altitudes. If this factor is calculated into carbon dioxide emissions, these rise to $400\,\mathrm{g\,km^{-1}}$.

The car industry is investigating numerous solutions to enable climate-neutral mobility in the medium and long term. These include:

- biofuels;
- electric cars using renewable electricity; and
- drive systems based on renewable methane, methanol, or hydrogen.

Although the natural gas vehicles currently being marketed emit slightly lower levels of carbon dioxide than before, they really do not offer an alternative for protecting the climate worldwide. Biofuels may be able to replace oil relatively easily and are sometimes added to conventional fuels. However, the problem is that not enough biomass is available to provide an adequate supply. Even if all the farmland in places like Germany and Britain were used to plant the raw materials needed for biofuels, this would not come close to covering the current fuel needs of motor vehicles in those countries. On closer inspection, even renewable hydrogen poses some problems. (Biofuels and hydrogen will be covered in detail in Chapters 12 and 13, respectively.)

Until now the main obstacle to electric cars has been limited battery life. Long charging times and the short distances that can be driven on a single charge are preventing the widespread use of these cars. In the meantime, however, enormous progress has been made, making electric cars an interesting alternative. The only way to achieve climate neutrality in transport by 2040 is to switch consistently to e-mobility. Assuming a service life of 15 years for cars, the last car with an internal combustion engine should actually roll off the production line in 2025. In order for e-mobility to offer a real alternative from a climate protection perspective, however, the electricity used to charge the batteries and manufacture the electric cars must come from renewable power plants.

3.3 Industry and Commerce – Everyone Else is to Blame

It is widely believed that industry and the energy companies are mostly to blame for greenhouse gas emissions and so private individuals cannot do much about the problem. But looking at the situation more closely, one sees that they are only complying with what the customers want. Therefore, ultimately it is the consumers who are responsible for the emissions because of the products they are demanding.

If all electricity customers were to switch to suppliers or tariffs that only offer electricity from renewable energies, the power supply would very quickly be carbon dioxide free. The only problem the energy suppliers would then have is being able to build enough renewable power plants as quickly as possible to accommodate this sudden demand.

Due to their choice of products, consumers are failing to exert pressure in the right places for a sustainable type of economy. The production of any type of product – whether a food item or consumer goods – requires energy and therefore causes carbon dioxide emissions. The higher one's personal consumption, the higher the use of energy and the higher the carbon dioxide emissions. Yet even those who completely renounce consumption will not be able to reduce their energy needs to zero, because high emissions are caused during practically all food production.

Each consumer can have an important impact on indirect energy consumption through product selection:

- Try to buy only high-quality and durable products.
- Give preference to regional products.
- Select products that require less energy during the manufacturing process and produce lower emissions.
- Give preference to companies with environmentally friendly policies.

3.4   Your Personal Carbon Dioxide Balance

Even if users follow all the energy-saving tips provided, they will probably find that the carbon dioxide emissions for which they are responsible are still quite high. This section enables readers to do a self-assessment by explaining how to calculate the carbon dioxide emissions an individual is personally responsible for.

3.4.1   Emissions Caused Directly by One's Own Activities

The easiest way to determine which emissions are the result of one's own activities is to look at how much oil, natural gas, petrol and electricity one has used. It is relatively simple to establish the amount of energy consumed per year by checking bills and invoices:

- annual electricity bill;
- heating bills;
- total distance travelled by car, and average fuel consumption;
- total distance travelled using public transport; and
- total distance travelled by plane.

Based on this information, the following calculation method can be used to determine the emissions resulting from one's own activities. The average values for Germany are highlighted and can be overwritten with your own values.

Specific consumption values may have to be adapted to a person's own particular circumstances. In electricity generation in Germany on average 0.566 kg of carbon dioxide is produced per kilowatt hour (kg CO_2/kWh) of electric power, and the value in the USA is roughly the same. Those who get green electricity from renewable power plants can cut their emissions from electricity consumption to zero. In Germany, all electricity suppliers are obliged to inform their customers about the composition of the electricity supplied and its environmental impact. The precise CO_2 emission values per kilowatt hour should therefore be available from the electricity supplier.

Emissions from heating can vary considerably. If heat is generated electrically, the emissions are the same as explained under electricity. Modern natural gas heating systems produce 0.2 kg CO_2/kWh of heat, modern oil heating systems 0.28 kg CO_2/kWh. With old, inefficient heating systems the emissions can rise from 0.25 to

0.35 kg CO_2/kWh. Heating with biomass at most generates indirect carbon dioxide emissions due to processing and transport. With wood pellets the value is around 0.06 kg CO_2/kWh.

The carbon dioxide emissions specific to a person's own car can be calculated on the basis of average fuel consumption *(see Planning Guide, p. 85)*. With air travel an emissions calculator makes it relatively easy to calculate emissions precisely, according to air route.

A. Calculation of Annual Private Direct Carbon Dioxide Emissions

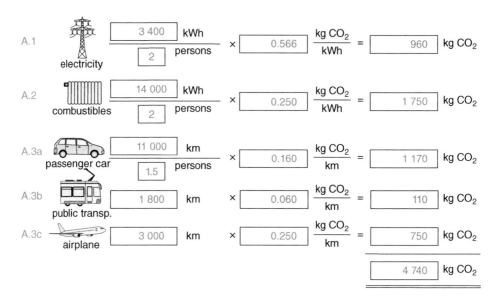

- www.atmosfair.de/en/   Atmosfair emissions calculator for flights
- www.qixxit.de   Comparison of means of transport
- www.greenmobility.de   Comparison of means of transport

3.4.2 Indirect Emissions

In addition to the emissions a person has caused directly, everyone is also indirectly responsible for other emissions. Energy is consumed in the production, processing, and transport of foodstuffs, consumer goods, and other products, and this in turn causes carbon dioxide emissions.

The public sector, including government offices, schools, the police, the fire brigade, as well as the departments responsible for highway maintenance, also need energy and therefore cause carbon dioxide emissions. About one-sixth of emissions are attributed

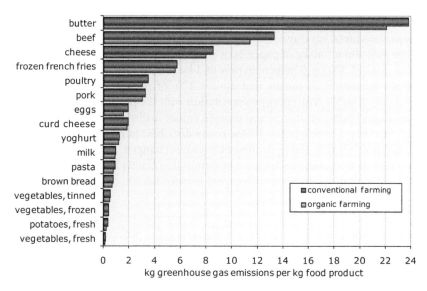

Figure 3.11 Greenhouse gas emissions converted to CO_2 equivalents for the production of different foodstuffs. Data: [Fri07].

to public consumption. This is an area where there are not many possibilities to make personal reductions.

About 1.3 t of carbon dioxide per person is produced every year due to our individual food consumption. Figure 3.11 compares the greenhouse gas emissions of different foodstuffs. In addition to carbon dioxide, the chart takes into account other greenhouse gases, such as methane and nitrous oxide, and converts these values into carbon dioxide equivalents according to how harmful they are to the environment. Beef, butter, and cheese do not come out favourably. Ruminant cows release large amounts of methane that have a high greenhouse effect. Subsequent freezing of the meat also adds to the energy toll, so that people who prefer frozen meat to fresh meat increase greenhouse gas emissions by another 10–30%.

The amount of processing that food requires also has a considerable effect on the climate. For example, fresh potatoes hardly tip the scales of greenhouse gas emission. However, the energy-intensive further processing required to make frozen French fries increases the greenhouse gas emissions contained per kilogram of food almost 30 times.

As women normally eat less than men and older people eat less than young people, clear differences exist from person to person. Going on a permanent diet is certainly not the right approach to reducing carbon dioxide emissions. However, people who buy seasonal products from local organic farms, reduce their consumption of meat, and largely avoid tinned and frozen foods can significantly reduce emissions resulting from food intake. Reductions of up to 30% can easily be achieved this way. However, those who need a large number of calories, because they do a lot of physical work or sport, and then obtain these calories mainly from frozen products and fast foods, could find they double their greenhouse gas emissions just from what they eat.

Personal consumption is responsible for around 2.9 t of carbon dioxide. This is mostly from the production, storage, and transport of all products that do not fall under the category of food supply. Included in this group are clothing, furniture, machines and equipment, paper products, cars, and housing.

Each German uses about 240 kg of paper each year, each Briton 200 kg or more, and each US citizen around 300 kg. The production of 1 kg of new paper produces about 1.06 kg of carbon dioxide. This means that almost 10% of personal consumption is attributed to the use of paper. The energy consumption with the production of recycled paper drops by around 60% and the carbon dioxide emissions by about 16%. Whereas recycled paper was very popular in the 1980s, many distributors no longer carry it at all – even though its benefits to the environment have not changed. The reason is weak demand on the part of the consumer because of unfounded claims that recycled paper destroys printers and copiers and is even damaging to one's health.

• www.papiernetz.de   Paper recycling initiative

The rule generally is that anyone who is a consumer is also contributing to high emissions. On average around 4–5 kg of carbon dioxide is created for each euro spent by consumers. Here, too there are many ways to make reductions. High-quality long-life products almost always produce low amounts of carbon dioxide. Although they are usually more expensive initially, in the long run they actually protect the wallet as well as the environment because of their long useful life. Increasing the use of natural materials may not always be less expensive, but it does usually reduce greenhouse gas emissions. For example, building a new solid house is almost as expensive as putting up a prefabricated house. However, if a wood frame construction is used for the prefabricated house, the additional use of wood as a building material will save around 20–30% in carbon dioxide emissions. As a result, the savings potential quickly runs into several tons.

In the following calculation method, the indirect greenhouse gas emissions can be inserted in the spaces shown. With the exception of public consumption, the average values are again highlighted in a lighter and can be changed individually upwards or downwards.

B. Calculation of Annual Indirect Carbon Dioxide Emissions

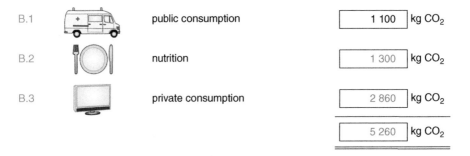

B.1	public consumption	1 100	kg CO_2
B.2	nutrition	1 300	kg CO_2
B.3	private consumption	2 860	kg CO_2
		5 260	kg CO_2

3.4.3 Total Emissions

Once the personal direct and indirect emissions have been determined, they can be added up to get a figure for all personal carbon dioxide emissions. At 9.6 t, almost 10 t of carbon dioxide per capita and year were still generated on average per person in Germany in 2015. In Great Britain the figure is 8.9 t and in the USA more than 19 t. If other greenhouse gases, such as methane and nitrous oxide, are taken into account and converted into carbon dioxide equivalents based on how damaging they are to the climate, then the total greenhouse gas emissions in Germany increases to more than 11 t per head per year. These results are, however, limited to carbon dioxide.

C. Calculation of Annual Total Personal Carbon Dioxide Emission

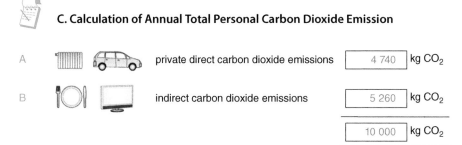

A	private direct carbon dioxide emissions	4 740	kg CO_2
B	indirect carbon dioxide emissions	5 260	kg CO_2
		10 000	kg CO_2

An even more precise calculation of personal emissions can be made using the emission calculators on the Internet. Figure 3.12 enables a comparison of one's own carbon dioxide emissions with the reduction targets and emissions of other countries. The colour scale in the figure shows how the emissions should be evaluated.

- http://uba.co2-rechner.de
- www.lfu.bayern.de/energie/co2_rechner

Online CO_2 calculators for Germany

3.5 The Sale of Ecological Indulgences

We may be able to make major reductions in our own direct and indirect emissions. However, in most industrialized countries it will still be relatively difficult to move

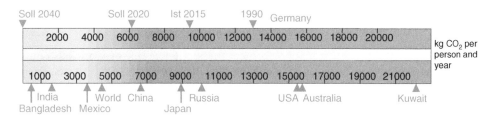

Figure 3.12 Scale of emissions of carbon dioxide per head and year.

emission levels into the 'green' zone. Some emissions, such as those related to public consumption, lie well beyond one's personal sphere of influence. Other reductions can only be achieved through radical changes in lifestyle or relatively high investment.

People who still want to make further reductions can use the emissions trade to compensate for their own emissions. This idea is even being practised on a large scale by states that want to reduce their overly high emissions as part of their commitment at the international level. The modern emissions trade is slightly reminiscent of the Medieval church's practice of selling indulgences. However, as the expression 'sale of indulgences' has negative connotations, the term joint implementation (JI) was invented.

Whatever is planned on a grand scale at the state level, and has partially already been implemented, can also be done in the private sphere. For instance, you could give your neighbour an inexpensive low-energy light bulb as a gift. This bulb could save up to 300 kg of carbon dioxide emissions over its lifetime. If you gave away enough low-energy light bulbs, you could conceivably save your total emissions in another area – at least theoretically. The actual personal emissions continue to accumulate and would then have to be transferred to the emission results of your neighbour. In pure calculation terms, the neighbour's emission results would remain constant. However, if the neighbour's emissions are also too high, he will not have an easy option for reducing his emissions himself. Furthermore, due to the savings in electricity costs, the neighbour will end up with more money in his household budget. If he then invests this money in a couple of litres of petrol for an extra jaunt in the car, then the whole exercise has ended up being highly counterproductive.

The situation is different if a low-energy light bulb is given to a school in a developing country with emissions in the green zone. In this case the gift of the light bulb will reduce already low emissions even further. Naturally the prerequisite even here is that the school is not in a position to buy the bulb itself. If the school is already planning to buy a low-energy bulb as an example of reducing electricity costs, the bulb given as a gift will not really represent a savings effect. Practised on a large scale, this type of investment is known as the Clean Development Mechanism (CDM).

- The following criteria should be included to ensure the success of climate protection projects:
- carbon dioxide reductions through financing renewable energy plants or energy-saving activities;
- implementation of an additional measure that would not necessarily have been carried out during the lifetime of a project;
- guarantee of successful project implementation and plant operation during the entire project; and
- ongoing development and technology transfer after project completion.

It is difficult for a private individual to meet these criteria. However, various companies offer professional services for implementing climate-protecting projects. Even major projects are possible if a large number of customers join forces. However, an independent auditing authority should always be available to monitor that the process is being followed correctly (Figure 3.13). Atmosfair is an example of a company that

board
of control

CO₂

investors   CO₂ credit   agent   CO₂ credit   reduction measure

Figure 3.13 Principle of private emissions trading.

offers a programme for offsetting emissions from plane trips. One of the projects in their programme involved replacing diesel stoves in large kitchens in India with solar mirrors.

Currently about €20–30 should be allocated to compensate for the emission of 1000 kg of carbon dioxide with clean development projects. Thus, it will cost a person €10 each month to prevent the emission of $5000\,kg\,yr^{-1}$. Protecting the climate is really not that expensive.

- www.atmosfair.de/en/
- www.co2ol.de/en/
- www.greenmiles.de

Various services for reducing carbon dioxide emissions through climate-protection projects

Naturally emissions can be offset through renewable energy projects at home. In countries such as Germany, for example, anyone who invests in a wind farm in a good location or installs a large photovoltaic system on an optimally orientated roof can even make a small profit. Germany's Renewable Energy Sources Act (EEG) makes this possible. This law establishes the subsidy levels for electricity from renewable energy suppliers, such as wind farms, photovoltaic systems and biomass, geothermal and hydropower plants. These subsidy levels are normally higher than regular subsidies for electricity from conventional power plants. The energy supply company into whose grid the plants feed their electricity is allowed to split the extra costs among all its electricity customers (Figure 3.14). This means that all customers – whether they want to or not – are helping to finance the renewable energy plants.

These power plant projects thus reduce the average emissions of electricity generation in Germany. However, they cannot be used to improve one's own carbon footprint. Although an investment in a renewable energy plant for electricity generation in Germany bears the entrepreneurial risk, the financing of a plant running according to plan is the responsibility of the general public. Anyone in Germany who wants to improve their own CO_2 balance and not that of the general public by building a renewable energy plant must invest in plants whose remuneration is not regulated by the Renewable Energy Sources Act.

However, this does not mean that most investments in renewable power plant projects in Germany would be pointless. On the contrary, the German Renewable Energy

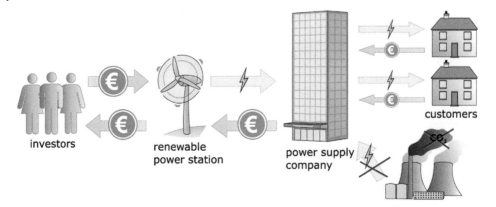

Figure 3.14 Principle of financing renewable power plants through the Renewable Energy Sources Act (EEG) in Germany.

Sources Act is the most successful instrument worldwide for promoting the rapid development of renewable energy. It aims to involve everyone equally in the social task of restructuring the energy sector – be it as investor, operator of renewable energy plants or financier for the reimbursement of electricity from existing plants.

4

'Energiewende' (Energy Transition) – The Way to a Better Future?

Just like sauerkraut, kindergarten or autobahn, the term Energiewende has the potential to become firmly established in the English language. It was coined in 1980, when the Freiburg-based Öko-Institut published a report entitled 'Energie-Wende: Growth and prosperity without oil and uranium'. At the time, however, the report met with a good deal of scepticism. The book was briefly reviewed in Die Zeit (weekly newspaper) and received the verdict 'dubious'. The term Energiewende only made its breakthrough after the reactor accident at Fukushima, Japan, in March 2011, which led to the German government's reversal of a previous decision to extend the operating life of German nuclear power plants and an announcement of a turnaround in energy policy.

It is true that the energy transition is closely linked to the nuclear phase-out. But the energy transition is not really completed until we have achieved a climate-friendly energy supply. The climate can only be saved if, in the medium term, all countries on Earth reduce their greenhouse gas emissions to almost zero. However, there are many among us who just cannot, or do not want to, imagine life without oil, natural gas and coal.

Yet, a mere 300 years ago, renewables made up the Earth's entire energy supply. It is quite certain that the world's energy supply will once again be completely carbon-free 200 years from now. By then the last deposits of fossil energy sources will have been exhausted. As a result of an almost 500-year history of fossil energy use, the climate would have totally collapsed by then. If we want to prevent this happening, we must convert to carbon-free energy supplies long before then. We only have about 30 years left to achieve this (Figure 4.1). It is therefore in the hands of our generation to either preserve or massively endanger the livelihoods of future generations.

4.1 Coal and Nuclear Power Plants – Crutch Instead of Bridge

A rapid energy transition is not easy to achieve, but this is not because it would be technically or economically impossible. Rather, the problem lies with numerous players who benefit greatly from today's energy supply structure and are among the losers in a rapid turnaround. Until a few years ago, many players who were supposed to implement the energy transition today, generally questioned its feasibility. Now, politicians and companies transformed from Saul to Paul are supposed to lead the most important

Renewable Energy and Climate Change, Second Edition. Volker Quaschning.
© 2020 John Wiley & Sons Ltd. Published 2020 by John Wiley & Sons Ltd.

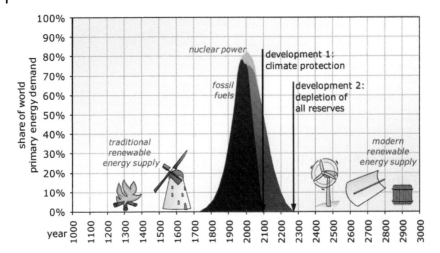

Figure 4.1 The energy transition won't be completed until we have established an energy supply system entirely without nuclear and fossil fuels. There are two scenarios for this: We can wait until energy supplies run out, or we can practise genuine climate protection.

project of our century for humankind's continued success on Earth. A healthy scepticism may be appropriate. Without people who maintain the necessary pressure, a quick turnaround is unlikely to succeed. Coal and nuclear power plants in particular, which are often referred to as bridges for the renewable energy age, are increasingly becoming a brake block, because their existence is increasingly hindering the rapid expansion of renewables.

4.1.1 Energy and Automotive Companies Have Bet on the Wrong Horse

Four major energy companies E.ON, RWE, EnBW, and Vattenfall controlled around 80% of the German electricity market for many years. In 2000 these companies had already agreed with the then 'red/green' federal government to phase out nuclear power. They obviously didn't take that agreement too seriously. True to the motto 'After red/green comes conservative/liberal and then we reverse everything', they worked behind the scenes to extend the operating life of German nuclear power plants and planned and built several new coal-fired power plants. In autumn 2010, this strategy was going to plan, with the extension of the operating lives of German nuclear power plants decided by the then conservative/liberal government, until Fukushima put a firm stop to it.

> **ⓘ The Last Dinosaur**
>
> This was the title of an award-winning report in Die Zeit (weekly newspaper) about Jürgen Großmann, the former CEO of RWE AG. From 2007 to mid-2012. He headed the fortunes of Germany's second-largest energy company. Since he was regarded as the driving force behind the decision to extend the operating lives of German nuclear power plants in 2010, he received the 'Dinosaur of the Year' anti-award from the Nature and Biodiversity Conservation Union. As the figurehead of the German energy industry, he is symptomatic

of the relationship between corporations and renewables, climate protection and coal and nuclear power plants with the following quotations:

12 July 2009: 'Germany is switching from coal to gas in order to be ecologically cleaner. … This is no way to protect the global climate'.

31 October 2009: 'Identical reactors are running in the Netherlands, France or Belgium for 60 years or more, in the USA there is even talk of 80 years. … With the current time limit of 32 years we remain below our economic possibilities'.

20 January 2010 'Imagine, 80% of our electricity generation depended on renewables: It's not just the lights that would go out in times like these'.

20 April 2011: 'Lignite is the German oil – this old RWE slogan applies in today's uncertain times more than ever'.

18 January 2012: The promotion of solar energy in Germany makes as much sense as 'cultivating pineapples in Alaska'.

Jürgen Großmann was replaced on 1 July 2012. In the meantime, the large German energy companies have restructured considerably. The language has become much more moderate. Business with fossil and nuclear power plants was partly outsourced to separate companies, and renewables are now officially very popular. Nevertheless, the large and climate-damaging power plants are still owned by the energy companies or the outsourced companies.

This corporate policy has wasted valuable years that could have been used to implement the energy transition. Despite the nuclear phase-out and climate protection promises, many energy companies and municipal utilities today still have a power plant fleet that is completely inappropriate for the energy transition. In 2012 RWE commissioned two new lignite-fired power plant units at Neurath, with an output of 2200 MW. Vattenfall's 1700-MW Moorburg coal-fired power plant in Hamburg was not finally connected to the grid until 2014. If these billion-euro investments are to pay off, the rapid phasing out of coal, which is urgently needed for climate protection, is practically impossible. This also explains why the expansion of photovoltaic (PV) and wind power plants in Germany has been significantly reduced in recent years and why Germany has virtually lost its pioneering role in the energy transition and in climate protection.

If nuclear and coal-fired power plants are to be quickly replaced by solar and wind power plants, they must be written off prematurely. This threatens companies with massive losses, which in extreme cases can even threaten their very existence (Figure 4.2).

Other industries are also blocking a rapid energy transition with undesirable technological developments. In the transport sector, for example, the energy transition can only succeed if the majority of vehicles are converted from internal combustion engines to efficient electric motors. The efficiency of electric motors is significantly higher than that of petrol or diesel engines, and the electricity can be generated without carbon dioxide emissions, using renewable power plants. What sounds like a look into a distant future today was already reality 100 years ago: About 40% of all cars in New York were electric at the time. At that time, the tedious cranking of petrol engines made them particularly unattractive. It was not until the invention of the starter motor in 1911 that the triumphant advance of internal combustion engines began. Today, car manufacturers find it difficult to break with a tradition that is more than 100 years old and to bring

Figure 4.2 Lignite-fired power plant Jänschwalde near Cottbus. The energy companies have the wrong power plant fleet for a quick energy transition.

modern electrical versions onto the market. Instead, they are intensively lobbying to delay stricter carbon dioxide limits and thus prolong the age of fossil engines. There may also be simple economic reasons to continue to apply the brakes: electric cars do not need oil changes and are less susceptible to wear. This is to the detriment of the attractive spare parts business and the workshops associated with repairs. In addition, batteries represent a major proportion of the value of electric cars. They have to be purchased from third parties and therefore do not make a significant contribution to the company's own added value. Finally, the main argument comes from the customers: despite climate change and rising fuel prices, less efficient but powerful cars continue to be trendy.

4.1.2 Lignite – A Climate Killer Made in Germany

Lignite-based power generation is one of the most climate-damaging forms of power generation. In 2016, Germany was the world's number one lignite producer, ahead of China.

Old lignite-fired power plants generate up to three times as much carbon dioxide as modern gas-fired power plants. Representatives of the electricity industry like to cite efficiency increases in the power plant sector as an important milestone towards effective climate protection. Indeed, enormous quantities of carbon dioxide can be saved if ailing, inefficient power plants are replaced with ultra-modern ones. Whereas old lignite-fired plants usually reached an efficiency of less than 40% in electricity production, new plants can easily reach 43%. Technically, efficiencies of 48–50% are possible.

For example, with an efficiency of around 35%, the Jänschwalde power plant near Cottbus in Germany is one the oldest and most inefficient large power plants in eastern Germany (Figure 4.3). With carbon dioxide emissions amounting to around 23.7 million tons in 2015, this plant alone produces about 3% of all carbon dioxide emissions caused by energy use in Germany.

Figure 4.3 Left: The village of Horno had to give way to opencast lignite mining in 2005. Right: After the coal has been mined, a 'lunar' landscape remains.

An example of the potential for increasing efficiency is the new BoA 2/3 Neurath dual power plant unit on the outskirts of Grevenbroich, Germany. BoA stands for 'Braunkohlekraftwerk mit optimierter Anlagentechnik' (lignite-fired power plant with optimized plant technology). The 2.6 billion euro plant with an efficiency of over 43% was commissioned in 2012, thereby saving around 6 million tons of carbon dioxide per year. The main problem, however, is that the power plant will continue to emit around 14 million tons of carbon dioxide every year. As an example, this is far higher than the total carbon dioxide emissions caused by the 40 million people in Kenya. With an average lifetime of 40 years, even a large, efficient power plant quickly develops into an obstacle to progress when it comes to effective climate protection. By the time the Neurath power plant is supposed to be disconnected from the grid in 2050, industrialized countries like Germany should have reduced their carbon dioxide emissions to zero. This is a level that not even the most efficient power plants can achieve.

The climate-damaging nature of lignite-fired power plants is not the only problem. They also release enormous amounts of toxic pollutants such as mercury, arsenic, or nitrogen oxides, thereby causing many health problems, including death. Enormous interference with nature and the landscape already occurs during opencast mining. Pumping off huge quantities of groundwater upsets the entire water balance of the region. After coal mining, a 'lunar' landscape remains (Figure 4.3), which requires considerable effort to renaturalize into a lakeland area (Figure 4.4).

Another problem with coal-fired power plants is that they are relatively difficult to regulate. Lignite-fired plants are usually designed for a base load. This means they work optimally when they deliver constant output for long periods of time. However, due to the increased use of wind power and solar energy plants, power fluctuations are increasing in the grid, and the demand for base-load power plants is decreasing accordingly.

To a small extent, lignite-fired power plants can compensate for power fluctuations. Reducing the output to 50% of the capacity is possible with new systems. However, this will not be enough for a rapid expansion of renewables. In spring 2012 there were already days when PV alone accounted for around 40% of the electricity supply in Germany at noon. On these days, PV already covered the entire midday peak. If the installed PV capacity continues to increase, it will increasingly replace base-load power plants. With

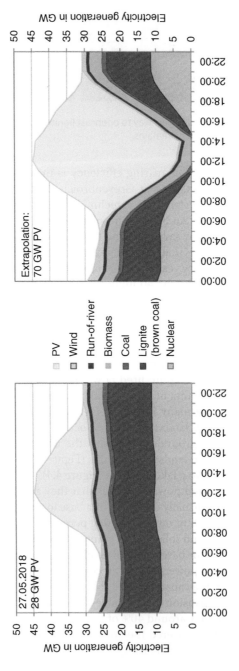

Figure 4.4 Left: Electricity generation from PV & wind power plants and generating units larger than 100 MW on 27 May 2012 in Germany with an installed PV capacity of 28 GW. Right: Same day with extrapolation to an installed PV capacity of 70 GW. Data: EEX Transparency.

an installed capacity of 70 GW, PV could cover almost the entire electricity demand at noon on individual days. This would then present the existing base-load power plants with almost unsolvable problems as once such plants have been taken off the grid, it takes hours for them to deliver their full power again. Reason enough for the energy suppliers and sympathetic politicians to demand a slower expansion of renewables in Germany and to throw all climate protection ambitions overboard.

4.1.3   Carbon Dioxide Sequestration – Out of Sight, Out of Mind

Since even a medium-term increase in efficiency in fossil power plants does not offer an alternative to the supply of climate-compatible energy, supporters of fossil energy plants have looked for arguments and ways to avoid coming under even more fire because of the damage caused to the climate.

Carbon dioxide sequestration is suggested as a way out of the dilemma. The idea behind it is simple and at first glance also appears highly plausible. The idea is that power plants of the future will no longer emit the carbon dioxide from burning coal and natural gas into the atmosphere, but instead capture and store it in a safe place (Figure 4.5).

The following options for the safe disposal of carbon dioxide are promising:

- manufacture of carbon-based building materials for the construction material industry;
- end-storage underground in depleted fossil deposits, in aquifers and in salt beds;
- compression or dissolution in the ocean; and
- bonding with special algae in the ocean.

However, a critical examination of the options available produces strong doubts about this method of carbon dioxide disposal. The technology for sequestration and storage is still at the research stage. When testing individual methods such as binding by algae in

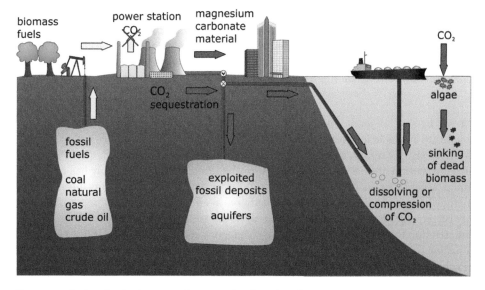

Figure 4.5 Options for final storage of separated carbon dioxide.

the sea, there have been severe setbacks which fundamentally call their suitability into question. Controversy also surrounds end-storage in the ocean. It is possible that the carbon dioxide will escape again into the atmosphere after a short period or will have an extreme but not yet identifiable effect on the ecological system of the oceans.

Underground storage locations are not available everywhere or would quickly be exhausted. If all power plants in the world were required to dispose of their carbon dioxide in a way that would not damage the climate, it would take an enormous logistical effort to transport the carbon dioxide over what could sometimes be thousands of kilometres to end-storage depositories. In addition, critics are concerned about risks from such deposits. If carbon dioxide were to escape uncontrolled in large concentrations, it could lead to death by suffocation of nearby humans and animals. In addition, numerous underground storage facilities are to be used for the storage of gas from renewables (cf. Chapter 13). If they were to be filled with carbon dioxide, they would no longer be available for renewables-related storage.

 Carbon Dioxide-Free Coal-Fired Power Plants – Not Really Free of Carbon Dioxide

With the development of technologies for the sequestration and storage of carbon dioxide, grandiose claims were made that power plants will one day be carbon dioxide-free. Strictly speaking, however, even carbon dioxide retention techniques can never make a fossil power plant free of carbon dioxide. The burning of fossil energy sources in these power plants inevitably produces carbon dioxide. So-called CO_2-free fossil power plants only separate carbon dioxide from other combustion gases so that it can be stored ultimately in a concentrated form outside the atmosphere. Carbon dioxide also escapes inadvertently during fuel processing, sequestration, and storage. In the long term up to 10% of carbon dioxide, and in individual cases even more, reaches the atmosphere again.

The term 'carbon dioxide-free coal-fired power plant' is therefore misleading. A company in the solar sector actually sued an energy supply company for misrepresentation over its claims that CO_2-free power plants were being built in Germany. The court ruled in favour of the solar company. Whether or not fossil power plants with carbon dioxide separation are carbon dioxide-free in legal terms is likely to be examined by other authorities.

It will be several years until proven technologies for capture and storage of carbon dioxide are available for commercial installations. Currently only studies and prototype facilities exist. Commercial introduction is envisaged for the year 2020 [Vat06]. Since the construction of final storage facilities for carbon dioxide in Germany has met with similar 'acceptance' as final storage facilities for nuclear waste, it remains questionable whether such facilities will even be built there. In addition, existing power plants can only be retrofitted at great expense, if at all. From today's perspective, carbon dioxide sequestration therefore appears more like a questionable attempt to legitimize climate-damaging fossil power plants.

Probably, the most important argument against global sequestration of carbon dioxide is the economic viability of doing so. Carbon dioxide sequestration generally reduces the efficiency of a power plant. Added to this is the cost of the transport and end storage of the carbon dioxide. It is difficult at present to estimate the exact cost involved. Many estimates predict that the cost increases for electricity from fossil power plants would be up to double present values [IPC05]. For many countries in the world this would rule out even the possibility of carbon dioxide sequestration. In contrast, cost estimates for renewable power plants show that in many locations they would prove to be more economical than fossil power plants within a relatively short period – regardless of whether or not they have carbon dioxide sequestration.

But perhaps in the future we will have to rely on carbon capture technology after all, in order to manage the consequences of climate change. Carbon dioxide could be absorbed through cultivation of biomass in the form of energy crops. If the biomass is then burned in power plants and the resulting carbon dioxide is separated and stored outside the atmosphere, it could even be possible to reduce the carbon dioxide content of the atmosphere again. However, it would make much more sense and be considerably cheaper to completely avoid the further increase in carbon dioxide concentrations through the use of renewables.

- http://www.ipcc.ch/pdf/special-   Background information on CO_2 capture
  reports/srccs/srccs_wholereport.pdf   and storage

4.1.4   Nuclear Power Comeback Was Not a Radiant Success

Almost every discussion on climate change ends up looking at nuclear energy as a possible saviour, despite the fact that its contribution to climate protection has always been small. However, after the nuclear accident at Fukushima, nuclear energy no longer seems to be an option for the majority, at least in Germany. The arguments against nuclear energy have already been explained in detail in previous chapters of this book. Even if the widely differing assessments of risks are ignored, nuclear energy does not offer an option for effective climate protection. The main reasons for this are:

- Nuclear energy only accounts for just under 5% of the world's primary energy supply, in Germany just over 6% in 2017, with a downward trend. In terms of final energy consumption, the proportion is even lower. Nuclear energy is therefore relatively insignificant for our energy supply.
- Electricity from new nuclear power plants is considerably more expensive than electricity from new solar and wind power plants.
- The amount of uranium that can be extracted economically is very limited. The price of uranium has already increased significantly in recent years.
- The output of nuclear power plants is only adjustable to a limited extent, which means they are not suitable for electricity supply systems with a high proportion of wind and solar power plants, whose output fluctuates by their very nature.
- Nuclear energy produces carbon dioxide only indirectly, e.g. during uranium mining. Greenhouse gas emissions are therefore considerably lower than with fossil power plants. However, nuclear energy carries other extremely high risks such as nuclear accidents, nuclear military conflicts or terrorist attacks.
- Plus, the thorny problem of permanent disposal of high-level radioactive waste has still not been resolved.

- The civilian use of nuclear energy also increases the risk of nuclear armed conflict through proliferation of uranium enrichment and nuclear fission technologies.
- Nuclear fusion will not be ready for use for several decades and thus too late to save the climate; it is also difficult to control and is extremely expensive.

ⓘ U-Turn Followed by U-Turn

On 14 June 2000, the 'red/green' federal government and the energy supply companies in Germany agreed for the first time to phase out nuclear energy. There was no fixed exit date. For all plants, residual electricity quantities were agreed that would probably have been reached in 2022.

On 28 October 2010, the conservative/liberal federal government decided to extend the operating lives of nuclear power plants by up to 14 years – a kind of 'exit from the nuclear exit'.

Due to pressure relating to the events of the reactor accident at Fukushima, on 6 June 2011 the conservative/liberal government decided to reinstate the decision to phase out nuclear power. Eight nuclear power plants were shut down immediately. For the others, fixed decommissioning dates apply. The last nuclear power plant is now scheduled to be shut down at the end of 2022. But once again, there are some voices from business and politics that want to reconsider this decision. To be continued?

Germany is not the first country to withdraw from the use of nuclear energy. Belgium and Switzerland have also decided to phase out nuclear power, but this will be completed later than in Germany. Italy shut down its four nuclear power plants as early as 1986. In a referendum in 2011, 94.1% rejected a comeback of nuclear energy. A referendum in 1978 also prevented the commissioning of the Zwentendorf nuclear power plant in Austria. The use of nuclear energy and the construction of nuclear reactors are now prohibited in Austria by a constitutional law.

4.2 Efficiency and CHP – A Good Double for Starters

4.2.1 Combined Heat and Power – Using Fuel Twice

When it comes to increasing efficiency, combined heat and power (CHP) is often promoted as a promising candidate. In conventional steam turbine power plants that burn lignite or hard coal, electricity is generated with an efficiency between 35% and 45%. Modern gas and steam turbine power plants, also referred to as combined-cycle gas turbine plants (CCGT), which use natural gas, can achieve up to 60%. However, this means that at least 40% of the primary energy is wasted through the cooling tower of the power plant.

CHP, or cogeneration, plants utilize the heat from electricity generation and are able to exploit up to 90% of the fuel. As a result, a well-designed plant produces less carbon dioxide than when electricity and heat are generated separately.

Many comparisons of CHP plants with plants that generate heat and electricity separately using fossil fuels show possible carbon dioxide savings of up to 50%. However, these comparisons usually pit modern CHP plants against antiquated electricity plants. If the comparison is made on a like-for-like basis, the carbon dioxide savings are reduced

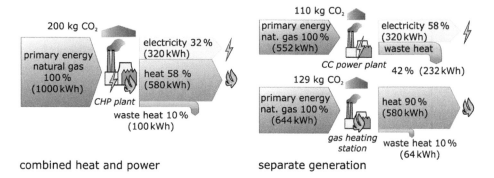

Figure 4.6 Comparison of primary energy demand and CO_2 emissions between CHP and separate generation of heat and electricity in modern plants.

to a meagre 15–20% (Figure 4.6) – too little to save the climate. Furthermore, these savings are only possible with optimal CHP plant operation. For example, in summer when a cogeneration plant is only supposed to generate electricity but not heat, it will have great difficulty in even coming close to the dream efficiency of 90%. A cogeneration plant can sometimes end up producing even more carbon dioxide than a straightforward electricity power plant.

If, on the other hand, sufficient heat demand exists over the entire year, cogeneration plants can help to reduce carbon dioxide. However, with CHP plants that use fossil fuels the savings for effective climate protection are too low. On the other hand, CHP plants that use energy sources such as biomass, renewably produced hydrogen, renewably produced methane or geothermal energy, are carbon-neutral and can accelerate the switch to a carbon-free energy supply.

4.2.2   Saving Energy – Achieving More with Less

As the previous chapter showed, the options for saving energy are enormous – at least in industrialized countries like Germany. The situation is different in developing countries. Someone who does not even own an electric light bulb and lives in a house without heating will be in no position to save energy through energy-saving bulbs or building insulation. Figure 4.7 shows that per capita energy requirements generally increase with the per capita gross domestic product (GDP), which roughly mirrors the prosperity of a country.

The differences between countries with the same GDP per inhabitant are considerable. Whereas the GDP per inhabitant in Canada and in Denmark is almost identical, a Canadian consumes far more than double the primary energy of someone living in Denmark. In addition to the way people live and how they handle energy, the climate conditions and industrial structure of a country also play an important role in energy consumption. However, having high energy needs does not automatically equate to high greenhouse gas emissions. Although the per capita energy requirement of Iceland is clearly higher than that of Canada, its carbon dioxide emissions are substantially lower. This is due to the fact that Iceland meets most of its energy requirements carbon-free, using hydropower and the natural heat of the Earth.

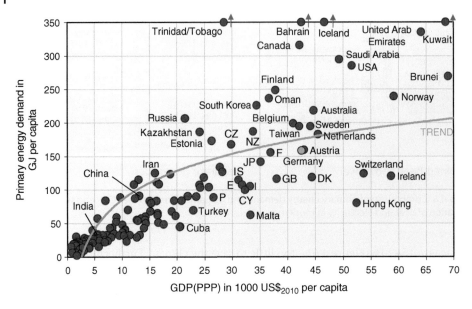

Figure 4.7 Per capita primary energy consumption based on gross domestic product (GDP) according to the purchasing power parity method (PPP) for different countries. Data: [IEA17], 2015.

Many countries in the world have a low level of prosperity and, consequently, a low demand for energy. Once the per capita energy requirements of China and India reach the same levels as Great Britain or Germany, this alone will more than double global demand. Considering the high growth rates of both countries, this would seem to be only a matter of time. If, in the interim, the industrialized countries were to halve their energy needs and the developing countries follow their example, this would considerably slow down the rise in energy demand.

Along with the growing per capita energy requirements in developing and emerging countries, the continuous rise in the world's population is also adding to the steady increase in energy demand. Between 1960 and 2000 the world population doubled, and energy requirements tripled (Figure 4.8). If the world population continues to climb at

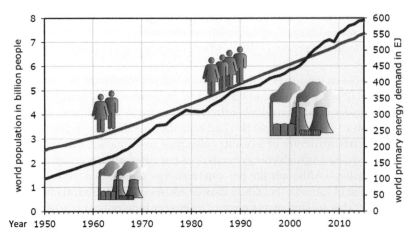

Figure 4.8 Development of worldwide primary energy demand and increase in world population.

this rate and reaches nine billion by 2050, this alone will mean a 50% increase in energy demand – even without any increase in per capita energy requirements.

These facts clearly show that energy-saving measures are enormously important in at least putting a brake on the increasing demand for energy worldwide. However, these measures alone will not be enough to achieve a major reduction in worldwide greenhouse gas emissions over the next 50 years. Along with implementing every conceivable measure to save energy, the most important thing will be to ensure that those energy requirements that cannot be eliminated are at least carbon-free. There is also a comprehensive solution to this: renewables.

4.3 Renewables – Energy Without End

The options for supplying climate-compatible energy discussed above offer only limited possibilities for reducing carbon dioxide. The situation is totally different with renewable energy sources: these offer almost unlimited potential.

Each year the sun radiates 1.5 quintillion kilowatt hours of energy towards the Earth. The atmosphere swallows up around 30% of this energy but over one quintillion kilowatt hours are still able to reach the Earth's surface. Our current primary energy needs are around 170 trillion kilowatt hours worldwide. By the way, a quintillion is a 1 with 18 zeros, and a trillion has 12 zeros. Therefore, the amount of energy that reaches the Earth's surface from the sun each year is 6000 times more than the total primary energy requirement of the world. So, we only need to use about one hour's worth of the solar energy that reaches the Earth's surface in order to cover the energy needs of the whole of mankind for a whole year.

Natural processes convert some of the sun's energy into other renewable forms of energy, such as wind, biomass, and hydropower. In addition to these energy forms, we are also able to use the natural heat of the Earth as well as tidal power derived from the motion of the moon in conjunction with other planets. All sources of renewable energy combined exceed the total fossil and nuclear fuels available on Earth many times over (Figure 4.9). In less than one day the sun radiates more energy to the Earth's surface than we could ever use if we were to burn all the oil reserves available on Earth.

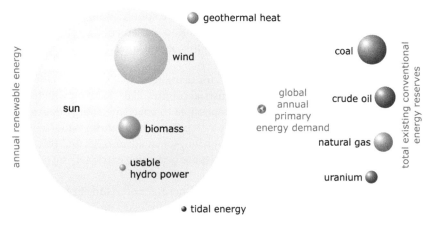

Figure 4.9 Comparison of annual renewable energy available and global primary energy requirement with the total existing conventional energy sources on Earth.

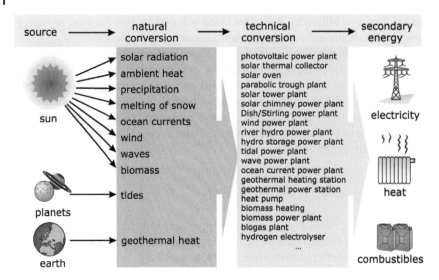

Figure 4.10 Sources of and possibilities for using renewables.

In discussions on climate, some critics question whether renewables are even able to cover our energy requirements. But a brief glance at the facts just mentioned shows that these doubts can safely be cast aside. The variety of possible uses of renewables is enormous. A number of different types of power plants can provide almost any amount of electricity, heating or fuel desired (Figure 4.10). The following chapters of this book will present in detail the key technologies for using renewables.

4.4 Germany Is Becoming Renewable

But first, we introduce a climate protection scenario for Germany to illustrate what a quick path for energy transition to a sustainable and carbon-dioxide-free energy supply could look like and what role renewables play in this. The responsibility that this entails for Germany is enormous. Many other industrialized countries are looking spellbound at Germany, wondering whether it will be possible to implement the fully-fledged energy transition without hiccups. If we succeed in this, it will spread to the rest of the world.

Compared to other countries, the possibilities for using renewables in Germany are far from optimal. However, if for a populous industrial country like Germany, which has only moderate renewable energy potential, it is possible to cover the energy demands entirely from renewables, this should certainly not pose a problem for other countries. Germany's leading role also offers numerous long-term advantages and this role is already paying off. Renewable energy technologies are becoming extremely successful exports for German industry.

However, the climate protection scenario outlined here is not a sure-fire scenario. The profiteers and supporters of the conventional energy industry are trying to delay the transformation of the energy supply. Cost arguments are used to prove that a rapid restructuring is unacceptable for the German economy and the population. No-one is denying that the transformation to a sustainable energy supply requires considerable investments. In the long term, however, a carbon-dioxide-free energy supply can save considerable costs that would have been spent on combatting the consequences of climate change and constantly increasing costs for fossil fuels. Due to the resistance from various quarters, the change is unfortunately not yet taking place at the required pace. However, an increase in the speed that this transformation takes would make it possible to achieve a carbon-dioxide-free energy supply in Germany by 2040, as the following climate protection scenario shows [Qua16].

4.4.1 All Sectors Are Important

If we want to successfully save the climate, we must reduce all greenhouse gas emissions to almost zero by 2040 at the latest. Around 84% of all greenhouse gases come from the energy sector from fossil fuel use. When it comes to energy, many only think of electricity, but this is not even responsible for half of Germany's climate gases. The rest is generated in the transport and heat sectors, as Figure 4.11 shows.

Sector Coupling

Just a few years ago, the term sector coupling was largely unknown. The term energy transition was essentially understood as the transformation of electricity supply to renewables. But for effective climate protection, greenhouse gas emissions from transport and heat supply must also be completely eliminated. In addition to the electricity sector, the heat and transport sectors must also be considered. Some authors also define the material use of fossil fuels and industrial processes as separate sectors.

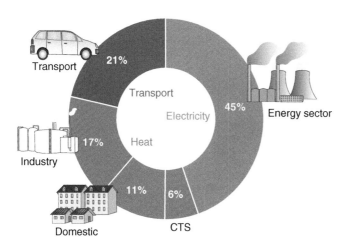

Figure 4.11 Share of different sectors in energy-related greenhouse gas emissions in Germany in 2015, CTS, commerce, trade, services. Data: [UBA17].

Today, the heat and transport sectors are dominated by climate-damaging oil and gas. The options for using biomass or geothermal energy in Germany are technically and economically limited. Therefore, in the future a large part of the energy for these sectors will have to be covered by electricity from solar and wind power plants. The electricity sector will thus be linked to the heat and transport sectors.

However, the expansion of renewables such as PV and wind power plants must be significantly increased through sector coupling, as this will considerably increase the electricity demand. Sector coupling offers enormous opportunities for the energy transition. Surplus from solar and wind power generation no longer has to be stored exclusively in expensive electricity storage facilities. When there is a lot of sun or wind, charging of electric cars can be given preference, in order to temporarily store electricity in the car batteries for times when there is little wind or sun. In the case of heat supply, excess electricity can also be used to heat water storage tanks or raise the building temperature. This facilitates a secure energy supply, saves costs and ensures technological developments that can also be exported to other countries in the future.

4.4.2 Energy Transition in the Heat Sector

It may be difficult to cut down on electricity consumption, but substantial reductions are possible when it comes to heating. The savings options in the building sector described in Chapter 3 enable a halving of the heat demand over the next 20–30 years. It is important that policy instruments are used to strongly promote modernization measures for existing buildings and that any new-build is in the zero-energy or plus-energy categories.

In 2015, 72% of the final energy consumption for space heating and 64% for hot water was still covered by fossil fuels (Figure 4.12). Among the renewables used directly for space heating and hot water generation, biomass dominated with a 12% share of final

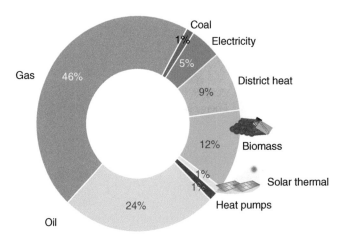

Figure 4.12 Share of different energy sources in final energy consumption for space heating and hot water in 2015.

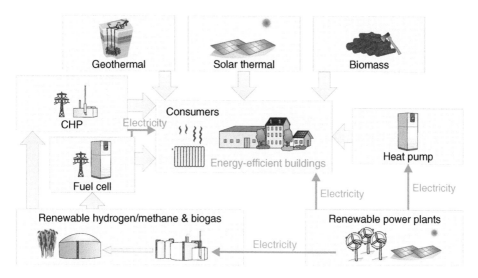

Figure 4.13 Components of a carbon dioxide-free renewable heat supply.

energy consumption. The share of solar thermal energy and heat pumps was comparatively insignificant at around 1% each. Deep geothermal energy was also used, but it is
statistically irrelevant.

However, the further expansion possibilities for biomass in Germany are limited. For
this reason, various renewables will ensure the supply of heat (Figure 4.13).

Compared to oil, natural gas performs better from an environmental and climate point
of view, but natural gas is also a fossil energy source that causes considerable greenhouse
gas emissions during combustion. A climate-friendly heat supply therefore requires
a departure from oil and gas heating. Some countries such as Denmark have already
decided to take such a step. CHP plants based on natural gas or oil are also unsuitable
for a climate-friendly heat supply.

Supporters of gas heating point out that in the future natural gas can be replaced by
climate-neutral gas from renewables and that natural gas only needs to be used temporarily. However, climate-neutral gas must first be produced in so-called power-to-gas
(P2G) plants in the form of hydrogen and methane by using electricity from renewable
power plants (Figure 4.14).

Losses in the order of 35% are to be expected. These losses occur in the form of waste
heat, some of which can be used, for example, to replace fossil-based district heating.
However, since P2G plants cannot run permanently, depending on the availability of
electricity from PV or wind power plants, the heat supply must then be secured by thermal storage.

If the remaining fossil energy demand is to be covered exclusively by gas condensing
boilers supplied by gas from P2G plants, an enormous additional electricity capacity
would be required to generate the required gas quantities. Figure 4.15 shows how the
electricity demand for single-family houses, which is currently in the range of 4000–
5000 kWh a^{-1}, would virtually explode if this path were chosen. The expansion of renewable power plants over a period of 20 years, which is what would be required, is totally
unrealistic.

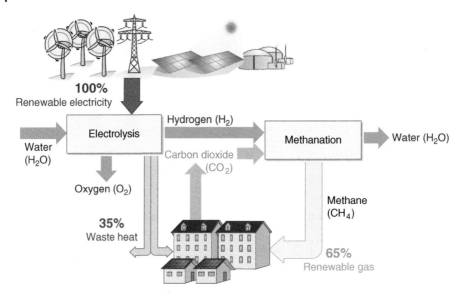

Figure 4.14 Principle of substitution of fossil-based natural gas by methane (power-to-gas, P2G) produced from renewable electricity [Qua16].

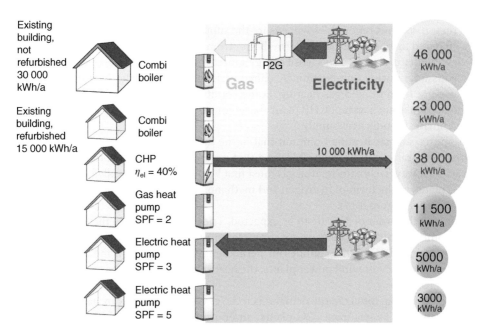

Figure 4.15 Efficiency and power consumption of electricity-based renewable heat supply systems [Qua16].

Heat pumps, on the other hand, will be able to achieve significant efficiency gains in the future. Heat pumps are electrical heating systems that use not only electrical energy but also low-temperature ambient heat. Modern heat pumps achieve seasonal performance factors (SPFs) of 3. This means that they can generate 3 kWh of heat from 1 kWh of electrical energy (see also Chapter 11). It is also conceivable to further increase the annual performance factor of heat pumps, e.g. through the use of waste heat recovery or solar thermal energy. Values of 5 can then be achieved. If heat pumps with an SPF of 5 are used and the heat demand of all buildings is halved through refurbishment, the electricity requirement for heating buildings is reduced by more than 90%.

Solar thermal systems are another option for renewable heat supply (see Chapter 6). However, these are still comparatively expensive at present. Nevertheless, it is conceivable that solar thermal energy could be integrated into local heating networks in larger units. This improves the economic efficiency considerably. If solar thermal energy is expanded more widely, the additional electricity demand in the heat sector can be reduced.

4.4.3 Energy Transition in the Transport Sector

At present, the biggest challenge is introducing climate-compatible changes in the transport sector. Here, the energy transition lags much further behind the heat sector. Only about 5% of the energy used in the transport sector in Germany comes from renewables. Most of these are biofuels such as biodiesel or ethanol, which are blended with conventional fuels. A significant increase in the proportion of biofuels is not possible because the agricultural land in Germany is not even close enough to cover the entire fuel demand with biomass fuels. In addition, part of the available biomass is required for electricity and heat generation. Biofuels should therefore be used primarily in transport sectors such as air or sea transport, where other alternatives are difficult to implement in the short term.

Other alternatives are needed for the vehicle sector. However, with broad support from German politicians, German car manufacturers have, in the past, repeatedly and faithfully resisted the implementation of climate protection measures. Some car bosses even claim that the savings being demanded would simply be physically impossible to implement. Yet, technically, modern cars are by no means energy-saving marvels. Even advanced combustion engines only achieve 25% average efficiency at best. This means at least 75% of the energy content dissipates unused, as waste heat into the environment.

Real efficiency gains can only be achieved by the electric motor, which easily achieves efficiencies of well over 80%. Figure 4.16 compares the efficiency of cars based on the range per kilowatt hour. A petrol-powered car with a fuel consumption of just over 7 l per 100 km, which corresponds to 65 kWh per 100 km, can only travel around 1.5 km with 1 kWh. If petrol is replaced by fuels such as hydrogen, methane, or methanol based on renewables, the range is reduced to 1 km kWh^{-1} due to the losses in the production of renewable fuels. If an efficient electric car is used instead of a car with an internal combustion engine, the range doubles. However, the electric motor cannot fully develop its potential because the fuel cells in the car first have to convert the renewable fuel back into electrical energy, which also results in losses. That is why a battery-powered electric car has the longest range. It eliminates losses during fuel production and reconversion to

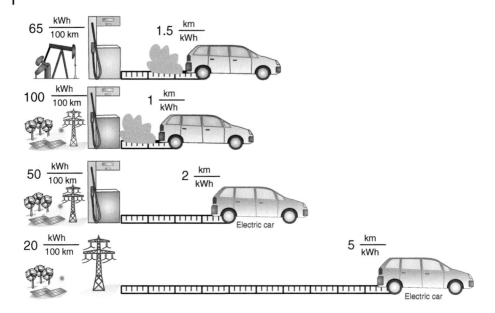

$65 \dfrac{kWh}{100\ km}$ $1.5 \dfrac{km}{kWh}$

$100 \dfrac{kWh}{100\ km}$ $1 \dfrac{km}{kWh}$

$50 \dfrac{kWh}{100\ km}$ $2 \dfrac{km}{kWh}$ Electric car

$20 \dfrac{kWh}{100\ km}$ $5 \dfrac{km}{kWh}$ Electric car

Figure 4.16 Comparison of the efficiency of different drive concepts for passenger cars.

electricity. The losses during charging and discharging of the battery are comparatively low.

The example shows quite clearly that a successful energy transition requires a consistent switch to electric cars. Thanks to technical innovations and massive cost reductions for vehicle batteries, battery-operated vehicles can already easily replace combustion engines in passenger cars and local goods transport. When longer ranges are required, a fuel cell vehicle using hydrogen from renewables can be an alternative. In long-distance freight transport, where long ranges are important, wire-bound electric trucks can replace the internal combustion engine. Such trucks have already been developed and are currently undergoing practical testing. The trucks are equipped with a hybrid drive and automatically latch onto an overhead line, when available. For large-scale operation it would make sense to electrify the inside lane of most motorways. Costs of around one million euros per kilometre are expected (Figure 4.17).

If we consistently focus on traffic avoidance, this would be of great help in implementing the energy transition in the transport sector. Therefore, public transport and the use of bicycles and footpaths should be significantly enhanced. For goods, an increased use of regional products is desirable. And if we succeed in bringing work, domestic, and leisure activities closer together in terms of space, traffic routes will also be reduced, and the quality of life will be improved.

4.4.4 Energy Transition in the Electricity Sector

In contrast to the heat and transport sectors, the transformation to a renewable electricity supply system has already started successfully. In 1990 hydropower, with its share

Figure 4.17 Electrified motorway with wire-bound electric truck. Source: Siemens, www.siemens.com/presse.

of around 3%, was the only renewable energy source worth mentioning. Over the next 20 years this was followed by wind power, biomass use and, more recently, PV. Their combined share was 17% in 2010, and by 2016 they had already reached 32%. However, carbon dioxide emissions did not fall by the same amount over the same period. While electricity was still imported into Germany in 2002, exports have increased enormously since then with declining electricity generation from nuclear power, which means that the share of coal-fired power generation has remained high.

If the use of oil and natural gas in the heat and transport sectors is also to be completely eliminated, electricity demand will increase significantly. Plus, there is the additional electricity consumption for a carbon-dioxide-free transformation of industry. If the electricity is then covered exclusively by renewables, it will mainly be based on fluctuating generators. To ensure a constant availability of the energy supply throughout the year, large storage capacities are required, consisting of battery and gas storage. This results in relatively large conversion and storage losses in the order of 20%. Figure 4.18 shows how this would increase electricity consumption.

In the future, electricity supply in Germany will become much more interesting and versatile. While it is virtually impossible to open up new potential for hydropower in Germany, wind power and PV have the greatest potential for expansion. In theory, both wind power and PV on their own could each cover the entire electricity demand in Germany. In practice, however, this does not make much sense, because large and expensive storage would be necessary due to the fluctuating availability of wind energy and solar radiation over the course of the year. However, a sensible combination of different renewables significantly reduces storage requirements. Due to their great potential, PV and wind power will cover the majority of the annual demand in the future. However, the construction of new PV and wind power plants will have to increase significantly in the coming years (Figure 4.19).

Controllable renewable power facilities such as geothermal plants or biomass power plants can only cover a small part of the supply in situations when the supply of wind and

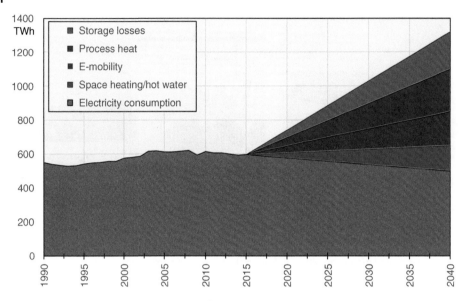

Figure 4.18 Increase in electricity demand if a climate-friendly energy supply is reached by 2040, taking into account sector coupling.

solar energy become simultaneously scarce. An intelligent control system must make sensible use of the various generators.

It is also conceivable that consumers could be switched on and off as required as part of smart grid concepts. Refrigerators, electric heat pumps or charging stations for electric cars could purchase more electricity from the grid than usual if there was a surplus of

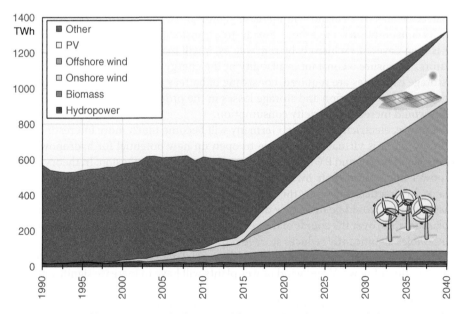

Figure 4.19 Possible expansion paths for renewables to reach a climate-neutral electricity supply.

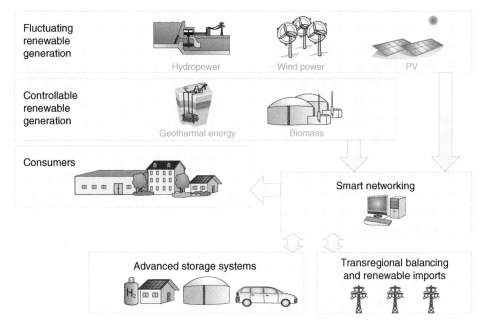

Figure 4.20 Components of a carbon dioxide-free renewable heat supply.

wind and solar energy. In the case of the refrigerator, the temperature would then drop slightly, in the case of the heating it would rise slightly, and in the case of electric cars the battery would be fully charged. If the supply subsequently falls again, these devices could curb their demand and then return to normal.

Further components of a sustainable electricity supply include balancing of the grid at a national level and import of renewable electricity. Despite all these measures, the storage requirements for a purely renewable electricity supply will rise sharply due to the large fluctuations in the supply of solar and wind energy. Batteries or pumped storage power Wplants are suitable for short-term storage. Hydrogen or renewably produced methane is an interesting energy source for storage over longer periods of time. The quantities required for this are relatively small compared to the possible demand in the heat or transport sector, and the associated losses are justifiable (Figure 4.20).

Fossil power plants and nuclear power plants are no longer necessary for a secure and climate-friendly electricity supply by the year 2040 at the latest and become an obstacle even before then. With the exception of gas-fired power plants, they cannot act as a bridge, as is often claimed. Lignite and nuclear power plants, in particular, are difficult to control and hinder smart operation in grids with high supply fluctuations. Gas-fired power plants, on the other hand, are much more responsive. Fossil-based, climate-damaging natural gas can ultimately be gradually replaced by biogas and renewably produced hydrogen or methane.

4.4.5 Reliable Supply Using Renewables

In purely arithmetical terms, renewables can thus cover the entire annual average energy demand. However, many people still question how electricity supply can be

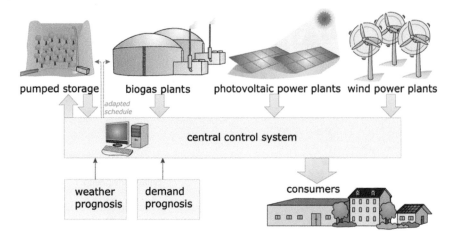

pumped storage | biogas plants | photovoltaic power plants | wind power plants

adapted schedule

central control system

weather prognosis | demand prognosis

consumers

Figure 4.21 Principle of a controlled combined cycle power plant for a reliable renewable power supply.

guaranteed – for example, if there is no wind blowing after the sun goes down. In this case, the output from PV and 78778 power plants would be zero. It is difficult to imagine that this would not cause massive problems.

With a balanced mix of different renewable power plants, which are intelligently linked through a central control system, the power supply can also be ensured in the case of highly fluctuating renewables (Figure 4.21). To prove this, various companies from the renewable energy sector initiated a Combined Power Plant project in 2007.

- www.kombikraftwerk.de

- www.unendlich-viel-energie.de

Information on the Combined Power Plant project
'Germany has unending energy' campaign

This Combined Power Plant linked and controlled 36 wind, solar, biomass and hydro-electric power plants scattered all over Germany. It covered exactly one ten thousandth of the whole German electricity demand. A central control system received information on the load profile and current weather forecasts. At times when wind and solar plants did not generate enough electricity, other plants had to step in. In the case of the Combined Power Plant, these were biogas plants and a pumped-storage hydropower plant. As biogas is easily stored, electricity can be generated at any time. Surplus supplies can then be stored temporarily in pumped-storage facilities or used to charge batteries for electric cars. There were very few instances when the wind and solar energy plants had to be cut back due to long sunny and windy periods. During these times a small proportion of the surplus was lost.

The results of the Combined Power Plant project were very promising. It turned out that on a small scale the demand could be covered almost optimally (Figure 4.22). The desired reliability of the electricity supply was achieved. So, there is no convincing

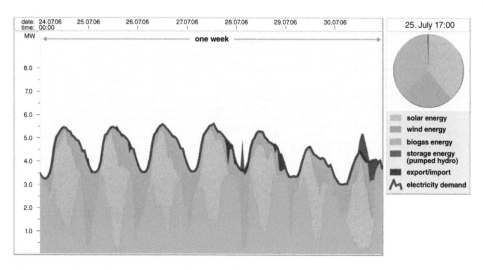

Figure 4.22 Shares of various renewable power plants in meeting energy requirements during a summer week in the Combined Power Plant project. Source: kombikraftwerk, www.kombikraftwerk.de.

reason why in the near future the entire energy supply cannot be completely covered by renewable energies in a similar way.

However, the storage requirements of a purely renewable electricity supply are several orders of magnitude higher than the capacities available today. At present, pumped-storage hydropower plants are mainly used for storage (see Chapter 9). However, the possibilities for constructing new pumped storage power plants in Germany are very limited. Therefore, the use of hydropower storage capacities in Norway or the Alpine countries is under discussion. Theoretically, there is much greater potential there than in Germany. However, the accessible capacities will also be significantly below the required storage requirements. The laying of the necessary power lines alone would cause major issues. High-voltage power lines across 'postcard fjords' are not very popular in Norway either.

New battery storage in PV systems or in electric cars could also contribute to meeting storage needs. However, even these capacities will not be sufficient for a secure, purely renewable electricity supply. Since, for effective climate protection, the extensive storage potential would have to be fully established within 30 years at the latest and only technologies that are quickly available on a large scale could be used.

The solution is seen in the natural gas network. Germany already has a very dense natural gas network with enormous underground storage facilities that can secure energy supplies for weeks. These networks and storage facilities can be used directly for the future electricity supply. The existing storage potential is already sufficient today to largely cover future demand.

Electrolysis (see Chapter 13) could produce hydrogen in the future from excess electricity at times when solar and wind output is high. Smaller quantities of hydrogen can be stored directly in the natural gas network. Up to a share of 5%, hydrogen can replace natural gas relatively easily. Shares of up to 20% are considered possible [Hüt10] (Figure 4.23).

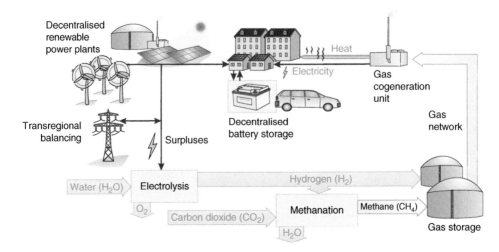

Figure 4.23 Use of the natural gas network to meet the storage needs of a purely renewable electricity supply.

Biogas can also be fed into the natural gas grid. If even higher hydrogen contents are to be achieved later, either the consumers can be converted to pure hydrogen operation or the hydrogen can be converted into renewable methane via methanation plants. The renewable methane could then directly replace fossil-based natural gas without further conversions. At times when there is little wind and solar energy available, the renewable hydrogen or methane is then fed back into the grid via cogeneration plants or fuel cells, which can also provide heat in addition to climate-friendly electricity.

4.4.6 Decentralized Instead of Centralized – Fewer Power Lines

'The construction of numerous new high-voltage lines is the central problem of the energy transition'; such statements proffer a message that is often repeated like a prayer wheel. However, the answer to the question of power line requirements is not entirely simple. In Radio Yerevan style it would be: In principle, power lines are helpful for the energy transition. But it depends on the structure that is envisaged for our renewable energy supply in the future. Maybe we don't actually need that many power lines.

A general distinction is made between distribution networks and transmission grids. The distribution networks transport the electricity to the end customers. In Germany, the distribution networks are often operated by regional utilities such as municipal utilities. There are now over 900 distribution network operators in Germany. The distribution network must be continuously expanded in order to be able to cope with electricity from new decentralized wind power and PV systems. In order to be able to charge a large number of electric cars in the future, the distribution networks must also be expanded. The replacement of local network transformers or the selective reinforcement of some existing lines is often sufficient for this purpose. In local networks, the lines are often underground, so that line reinforcement usually has a high level of acceptance among the population.

Figure 4.24 High-voltage lines are among the most controversial elements of the energy transition. In the long term, they are needed for the connection of offshore windfarms, for example. In the short term, they are mainly used for the distribution of surplus electricity from poorly controllable conventional power plants.

The situation is different for transmission grids. They consist of extra-high voltage lines, the so-called electricity highways, which transport electricity over long distances (Figure 4.24). These are usually large overhead lines, which often meet with little approval from local residents. For a strong expansion of offshore wind energy use at sea or the construction of large capacities of wind and solar power plants in sparsely populated regions, these lines must be strengthened, as their electricity cannot be consumed locally. The continued operation of inflexible lignite or nuclear power plants also increases power line demand. If conventional power plants continue to operate at times of high solar and wind supply because they are difficult to control, there will be regional surpluses that will have to be removed via the extra-high voltage lines. At times of high solar or wind supply, Germany now regularly exports large quantities of electricity abroad.

A power supply based entirely on renewables will not function without the use of offshore wind energy and thus without new transmission grids. However, the decentralized expansion of solar and wind power is much more important for a fast and cost-effective energy transition. If solar and wind power plants were first installed regionally where the electricity is also consumed, the expansion of the transmission grids could be significantly reduced. In a decentralized supply structure, gas-based storage facilities and rapidly controllable reserve power plants have a higher priority. They can quickly replace coal-fired and nuclear power plants with renewables.

4.5 Not So Expensive – The Myth of Unaffordability

The central message from politicians and energy companies that a rapid energy transition is unaffordable seems to have been their mantra in recent years and it is true that electricity prices more than doubled between 2000 and 2017. Figure 4.25 shows the development of electricity costs in Germany, which comprise the generation of

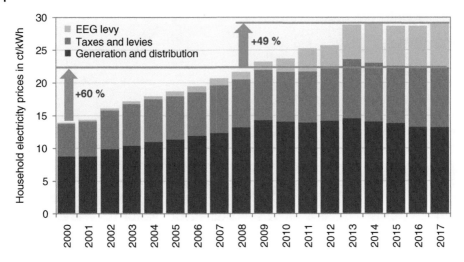

Figure 4.25 Development and composition of household electricity prices in Germany.

conventional electricity and distribution, taxes and duties as well as the so-called EEG **(Erneuerbare-Energien-Gesetz)** levy. The EEG levy covers the additional costs of generating from renewables compared to conventional power plants, and is charged to households and smaller commercial customers. It has fallen into disrepute as the cause of the electricity price increases. But even if we completely abolished the EEG levy, electricity prices would still have risen by a substantial 60% between 2000 and 2017.

The EEG levy is actually higher than it has to be. Many industrial companies are exempt from the levy. Plus, the increasing capacities of renewable power plants are decreasing prices on the electricity exchanges. This is good for industrial customers who purchase cheap electricity directly on the stock exchange. Household customers, on the other hand, do not like these price reductions. Since the EEG levy is calculated from the additional costs of renewables compared to the stock exchange electricity price, falling stock exchange electricity prices result in a higher EEG levy and thus higher household electricity prices.

Falling coal prices are a disaster for achieving climate protection targets. The CO_2 certificate trading should actually increase the price of electricity from climate-damaging power plants and thus lead to a reduction in demand and hence emissions. However, the economic crisis in Europe, far too generous allocation of allowances and the rapid expansion of renewables have led to an enormous oversupply of allowances and thus to a dramatic drop in prices. However, the consequential costs of unbridled carbon dioxide emissions will also have to be paid in the future. There are no reserves for this. The Federal Environment Agency puts the actual consequential costs of climate change at €80 t^{-1} of carbon dioxide [UBA14]. At the beginning of 2018, the price of CO_2 certificates was around €8 t^{-1} of carbon dioxide. In Germany alone, the unallocated consequential climate costs thus correspond to a subsidy of well over 20 billion euros for fossil power plants.

A study by the Fraunhofer Institute for Solar Energy Systems clearly shows that a completely renewable energy system will not be more expensive than the current supply system [ISE12]. However, this only applies to the final expansion phase. The investments

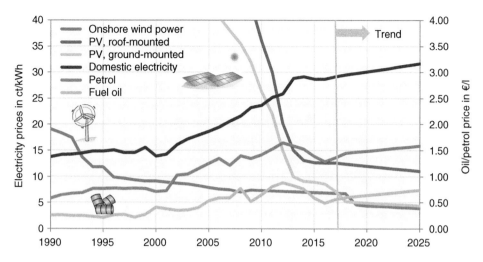

Figure 4.26 Development of prices for household electricity, heating oil, petrol, generation of PV electricity and wind power in Germany.

for the transformation inevitably lead to higher costs during the transition period. This is not really surprising. Electricity prices also had to be increased when nuclear energy was introduced. While the prices for renewables are falling continuously, the overall trend for fossil fuel prices, despite some ups and downs, is up. The prices for heating oil roughly tripled between 2000 and 2013. In the same period, the price of solar power fell to a quarter (Figure 4.26). And this trend is continuing. Even today, renewables are already competitive in many areas of the world, without subsidies, compared to new conventional power plants. The countries that first decoupled from oil, coal, natural gas, and uranium and their associated price increases will therefore also benefit significantly financially.

4.6 Energy Revolution Instead of Half-Hearted Energy Transition

4.6.1 German Energy Policy – In the Shadow of Corporations

A major obstacle to a rapid and climate-friendly transformation of our energy supply is the close intertwining of energy and automotive companies and politics. Before or after their active careers, many politicians held high positions in energy and automotive companies or worked as consultants. As a result, the energy transition is always designed with the interests of corporations at the forefront, which have placed great emphasis on fossil fuels as well as petrol and diesel vehicles. Effective climate protection and the interests of the population are often put at the bottom of the list.

Due to this intertwining, more and more contradictory decisions are made, which make a quick and meaningful energy transition more difficult and cause unnecessary additional costs. Figure 4.27 clearly shows that in 27 years no federal government in Germany has succeeded in reducing the share of fossil power plants in electricity generation. Initially, the increase in electricity generation from renewables compensated for

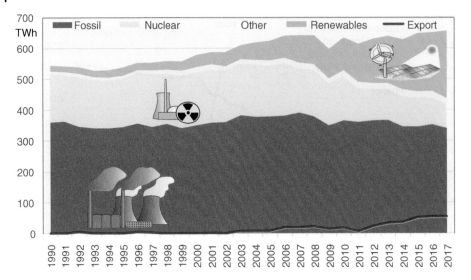

Figure 4.27 Development of gross electricity generation in Germany and electricity exports. Despite the energy transition, the proportion of fossil-based power plants remains largely constant. Data: [AGEB12].

the decline in nuclear power plant electricity and the increase in electricity consumption. Electricity exports were then noticeably increased. In 2017, German power plants were already able to cover three quarters of Austria's electricity demand in purely arithmetical terms.

Although there have also been low carbon dioxide savings in recent years, due to increases in the efficiency of fossil-based power plants, reductions of a completely different magnitude are, however, required for effective climate protection. This can only be achieved through a rapid reduction in the proportion of fossil-based power plants and a significantly faster expansion of renewables. Instead, the expansion of PV systems in Germany was reduced from 7.6 GW in 2012 to 1.5 GW in 2016, resulting in the loss of around 80 000 jobs in the PV sector. A significant reduction in wind energy expansion was then initiated for 2019, even though Germany has thus lost all chances of achieving the climate protection targets it had set itself for 2020. The only discernible goal of the political restrictions on supplies was to prevent a sharp drop in fossil-based power generation.

4.6.2   Energy Transition in the Hands of the Citizens – A Revolution Is Imminent

Politicians and energy companies are at best very cautious in supporting the energy transition. The driving force is the population itself. Over 40% of the 100 GW of renewable electricity generation plants in 2017 belonged to private individuals and farmers.

Only 5.4% were owned by the four major energy supply companies (Figure 4.28). As a result, they have noticeably lost influence and market share in recent years. With their influence on politics, they have played a major role in slowing down the energy transition in Germany. They have tried to adapt to the new market environment with

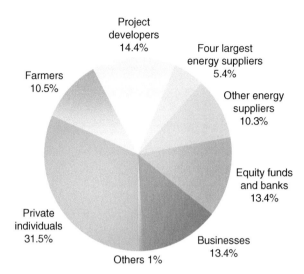

Figure 4.28 Distribution of the ownership of the renewable energy plants that provide Germany's electricity. Data: AEE/trend research, 2017.

restructuring measures. However, due to the large number of conventional power plants still in operation, they are only able to make slow adjustments.

But not to worry; as in the past, many other players will continue to drive forward the energy transition in the future. This means that the democratization of our energy supply is in full swing. A real energy transition is emerging where energy companies will play an ever decreasing role. Everyone can participate in the energy transition by investing in renewables. Those who do not have their own roof for the construction of a solar plant can participate in one of the numerous new operating companies or energy cooperatives and thus also benefit from the energy transition. In future, profits will no longer be limited to large energy companies. This means that it is in our own hands to quickly create a sustainable energy supply, so that we can also preserve the lives of future generations. It is up to us now.

5

Photovoltaics – Energy from Sand

The term 'photovoltaics' comes from two words, 'photo' and 'Volta'; photo stands for light and comes from the Greek phõs, or photós. The Italian physicist Alessandro Giuseppe Antonio Anastasio Count Volta, who was born in 1745, was the inventor of the battery and together with Luigi Galvani is considered to have discovered electricity. There is not much that associates him with photovoltaics. However, in 1897, 70 years after Volta's death, the measurement unit for electric voltage was named volt in his honour. Photovoltaics, or PV for short, therefore stands for the direct conversion of sunlight into electricity.

While experimenting with electro-chemical batteries with zinc and platinum electrodes, the 19-year-old Frenchman Alexandre Edmond Becquerel found that the electric voltage increased when he shone a light on them. In 1876 this phenomenon was also proven with the semiconductor selenium. In 1883 the American Charles Fritts produced a selenium solar cell. Due to the high prices of selenium and manufacturing difficulties, this cell was not ultimately used to produce electricity. At the time, the physical reason why certain materials produce electric voltage when radiated with sunlight was not understood. It was not until many years later that Albert Einstein was able to specify the photo effect that causes this. He eventually received the Nobel Prize for this work in 1921.

The age of semiconductor technology began in the mid-1950s. The abundant semiconductor material *(see info box in the next section.)* silicon became all the rage in technology, and in 1954 the first silicon solar cell made its appearance at American Bell Laboratories. This was the basis for the successful and commercial further development of PV.

5.1 Structure and Function

5.1.1 Electrons, Holes, and Space-Charge Regions

Understanding the relatively complicated way that solar cells work requires immersion into the most extreme depths of physics, but the small applied model shown in Figure 5.1 roughly explains the principle involved. There are two horizontal levels. The second level is located slightly higher than the first one. The first level has a large number of small hollows filled to the top with water. The water here cannot move by itself. Now someone starts to throw small rubber balls at the first level. If a ball hits a hole, the water

Renewable Energy and Climate Change, Second Edition. Volker Quaschning.
© 2020 John Wiley & Sons Ltd. Published 2020 by John Wiley & Sons Ltd.

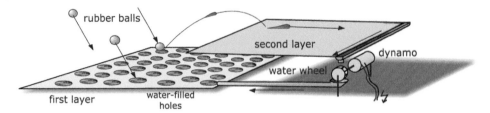

Figure 5.1 Model illustrating the processes of a solar cell.

splashes upwards and ends up on the second level. Here there are no hollows to contain the water. The second level is therefore inclined so that the water runs off and reaches the draining groove on its own. This groove is connected to the second level through a pipe and as the water flows through, it drives a small waterwheel with a dynamo. When the water reaches the lower level, it fills up the hollows again. The cycle can start all over again with new rubber balls.

However, we want to use solar cells not to produce a water cycle but to generate electric current to run electrical appliances. Electric current is created from the flow of negative charge carriers, called electrons. These are the same as the water in our simple model. The solar cell needs a material with two levels: one level where the electrons are firmly fixed, like the water collecting in the hollows, and a second level where the electrons are able to move freely. Semiconductor materials normally have precisely these properties. Tiny particles of light, called photons in physics, correspond to the rubber balls and are able to raise the electrons to the second level.

Conductors, Non-conductors, and Semiconductors

Conductors such as copper always conduct electric current relatively well, but non-conductors such as plastics conduct almost no electricity at all. In contrast, semiconductors – as the name indicates – only conduct electric current sometimes, for example at high temperatures, when fed with electric voltage or when radiated with light. These effects are used in the production of electronic switches like transistors, computer chips, special sensors and even solar cells.

Organic semiconductors are available in addition to elementary semiconductors, such as silicon (Si), and compound semiconductors, such as gallium arsenide (GaAs), Cadmium telluride (CdTe), and copper indium diselenide ($CuInSe_2$). All these materials are used in PV.

The tilt in our simple model is important, because it enables the cycle to function perfectly. Otherwise the water will not collect on its own in the rain gutter. With semiconductors the second level must also have an incline that enables the electrons to gather on one side. In contrast to our simple model, it is not gravity that is used to collect the electrons but instead an electric field, which pulls the negatively charged electrons to one side. To produce this field, the semiconductor is 'doped'. One side of the semiconductor is deliberately contaminated with an element like boron and the other side with a different element like phosphorus. As boron and phosphorus also have a varying number of

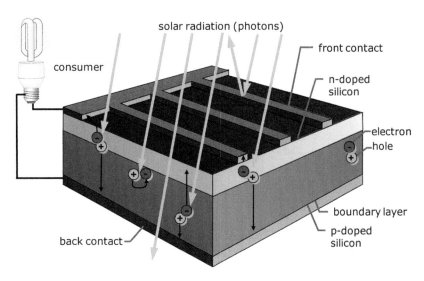

solar radiation (photons)

front contact

n-doped
silicon

consumer

electron
hole

back contact

boundary layer

p-doped
silicon

Figure 5.2   Structure and processes of a solar cell [Qua13].

electrons, they produce the necessary incline. The crossing area is called a space-charge region. An electric field is created here, which pulls the electrons to one side. There, external contacts collect them and they flow back to the first level through an external electric circuit. In the process they produce electrical energy.

Figure 5.2 shows the principal structure of a silicon solar cell. In technical jargon the different doped sides of the silicon wafer are called n-doped and p-doped silicon, respectively. Between the two areas is a barrier layer with the space-charge region. Light in the form of photons separates negatively charged particles (electrons) and positively charged particles (holes) and ensures that the electrons are able to move about freely at the second level. In contrast to the simple model shown, the holes also move. The electrons and holes are separated by the space-charge region. Thin front contacts collect the electrons on the front side of the cell.

However, not every light particle ensures that an electron is separated from the hole. If the energy of the photon is too low, the electron will fall back into the hole. On the other hand, if the energy of the photon is too high, only part of it is used to separate the electron from the hole. Some photons also move through the solar cell unused; others are reflected by the front contacts.

5.1.2   Efficiency, Characteristics, and MPP

The efficiency of the solar cell describes what proportion of the solar radiation power the cell converts into electrical power.

Solar Cell Efficiency

$$\eta = \frac{P_{el}}{\Phi} \left(efficiency = \frac{electrical\ power\ output}{incoming\ solar\ radiant\ power} \right)$$

Table 5.1 Efficiencies of different solar cells

Cell material	Max. cell efficiency (lab.) (%)	Max. cell efficiency (mass prod.) (%)	Typical module efficiency (%)	Surface required for 1 kW$_p$ (m^2)
Monocrystalline silicon	25.8	24	19	5.3
Polycrystalline silicon	22.3	20	17	5.9
Amorphous silicon	14.0	8	6	16.7
CIS/CIGS	22.6	16	15	6.7
CdTe	22.1	17	16	6.3
Concentrator cell	46.0	40	30	3.3

The higher the efficiency, the more electric power the solar cell can generate per square metre. In addition to the type of materials used, the quality of the manufacturing also plays a major role. Today silicon cells in mass production reach a maximum efficiency of almost 24%. More than to 25% efficiency has already been reached in laboratories (Table 5.1).

Incidentally, conventional petrol engines do not reach a level of efficiency higher than silicon cells. Compared to the 5% efficiency of the first solar cells in 1954, technology has advanced considerably. If individual solar cells are packaged into PV modules, the efficiency drops somewhat due to the space necessary between the cells and the module frames. It is hoped that, in the future, other materials can be used to produce further cost savings. Compared to silicon cells, however, the efficiencies still need to be improved. Concentrator cells in which sunlight is concentrated through mirrors or lenses reach a very high efficiency but are considerably more expensive than normal silicon cells.

In addition to efficiency, there are other parameters that describe PV modules. Data sheets for PV modules usually contain a current-voltage characteristic curve. The maximum current I_K flows with a short-circuited PV module. The short-circuiting is harmless for the module. The short-circuit current is limited and depends on the solar radiation intensity, also referred to as irradiance. If nothing is connected to the PV module, it is in open-circuit state and no current flows. In this case it adjusts itself to the open-circuit voltage V_{OC}. A PV module is unable to produce any power when in short-circuit or open-circuit state. Between open-circuit and short-circuit state, the current depends on the voltage. The basic characteristic curve is similar for all solar modules (Figure 5.3).

In practice, the aim is to extract the maximum power from the PV module. This corresponds to the largest rectangle that can be fitted under the characteristic curve. The upper right edge of the rectangle on the characteristic curve is called maximum power point or MPP. The corresponding voltage is called MPP voltage or V_{MPP} for short. At this voltage, the PV module delivers the maximum power. In practice, operation near the MPP can be achieved by connecting a battery whose voltage is close to the MPP voltage, for example, or by using an inverter that automatically adjusts the MPP voltage on the PV module.

The current of PV module, and hence the power, decreases according to the number of incoming photons, i.e. the irradiance of the sunlight. For example, if the solar irradiance drops by 50%, the output of the PV module also drops by 50%. The output of PV

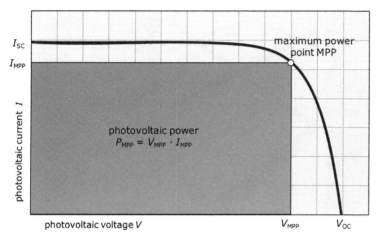

Figure 5.3   Current-voltage characteristic curve of a PV module.

modules also drops at high temperatures. If the temperature rises by 25 °C, the output of crystalline solar cells drops by around 10%. Therefore, when installing PV modules, care should be taken to ensure they are always well-ventilated and a draft of air cools the modules.

Standard test conditions (STCs) for comparing PV modules have been agreed internationally. The MPP power of solar cells and modules is determined at a solar irradiance of 1000 W per m^2 and a module temperature of 25 °C. Since in practice the irradiance is usually lower and PV modules can heat up to over 60 °C in summer, the MPP power determined under STCs represents a maximum value. This value is only achieved rarely and exceeded even less frequently. For this reason, the unit 'Watt Peak', W$_p$ for short, is also used (Table 5.2).

5.2   Production of Solar Cells – From Sand to Cell

5.2.1   Silicon Solar Cells – Power from Sand

Silicon, the raw material for computer chips and solar cells, is the second most common element in the Earth's crust after oxygen. However, in nature silicon occurs almost

Table 5.2   Important parameters for PV modules

Parameter	Symbol	Unit	Description
Open-circuit voltage	V_{OC}	Volt, V	Voltage of a PV module in open-circuit state without connected load
Short-circuit current	I_{SC}	Ampere, A	Current of a PV module in short-circuit state
MPP voltage	V_{MPP}	Volt, V	Voltage at which a PV module delivers maximum power
MPP current	I_{MPP}	Ampere, A	Current associated with the MPP voltage
MPP power	P_{MPP}	Watt, W	Maximum power a PV module can deliver

Figure 5.4 Polycrystalline silicon for solar cells. Left: Raw silicon. Centre: Silicon blocks. Right: Silicon wafers. Source: Photos: PV Crystalox Solar plc.

exclusively as an inclusion in quartz sand or silicate rock, or as silicic acid in the world's oceans. Even the human body contains around 20 mg of silicon per kilogram of body weight.

Pure silicon, on the other hand, is usually obtained from quartz sand. Chemically, quartz sand is pure silicon dioxide (SiO_2). To obtain silicon from it, the oxygen atoms (O_2) have to be separated at high temperatures. This process is called reduction and is carried out in arc furnaces, for example, at temperatures of around 2000 °C. The result is industrial raw silicon with a purity of 98–99%.

Raw silicon has to be purified further before it can be used to produce solar cells. The Siemens process is usually used for this purpose. Hydrogen chloride is used to convert the raw silicon into trichlorosilane, which is then distilled. At high temperatures of 1000–1200 °C the silicon is then separated again into long rods. The resulting polycrystalline solar-grade silicon has a purity of over 99.99% (Figure 5.4).

The silicon is melted down again to produce semiconductor silicon for computer chips and monocrystalline solar cells. In the crucible process invented by Polish chemist Jan Czochralski, a crystal seed is dipped into a crucible with a silicon melt and then slowly pulled upwards in a rotating movement. The molten silicon attaches itself to the crystal and a long, round silicon rod is created. In the process the silicon crystals align in one direction. This creates monocrystalline silicon. Most of the impurities remain in the melting crucible so that the semiconductor silicon is left with purities of over 99.9999%.

In the next step, wire saws cut the long silicon rods into thin slices, called wafers. This sawing process results in significant waste, with up to 50% of the valuable silicon material being lost as a result. The alternative is for two thin wires to be pulled through the liquid silicon melt. With this procedure, thin silicon wafers are formed between the two wires. Immersion in acid will remove sawing damage from wafers and smooth the surfaces. Several years ago, silicon wafers had a thickness of 0.3–0.4 mm. Nowadays, in order to save material and costs, the wafer thickness is reduced to well below 0.2 mm. This used to be a major technical challenge, since the ultra-thin wafers must not break apart.

Figure 5.5 Polycrystalline solar cells with anti-reflective coating before the front contacts are applied. Source: Photo: BSW, www.sunways.de.

The finished wafers are exposed to gaseous doping substances referred to as dopants. This produces the p- and n-layers described earlier. A transparent anti-reflection layer of silicon nitride less than a millionth of a millimetre thick gives the silicon solar cell its typical dark blue colour. This layer reduces the reflection loss of the silver-grey silicon on the front side of the solar cell. The darker the cell appears, the less light it reflects (Figure 5.5).

The front and back contacts are then applied using screen printing. To reduce the losses at the opaque front contacts, some manufacturers conceal them under the surface or try to move them to the back of the cell. Although this increases the efficiency of cells, it is also a more complicated and expensive way to manufacture them. The finished cells are then tested and sorted according to performance classes for further processing into PV modules.

5.2.2   From Cell to Module

Silicon solar cells are usually square in shape. The length of the edge is measured in inches. Originally, solar cells were typically 4 in. (approx. 10 cm) in length. Meanwhile, a measurement of 6 in. (approx. 15 cm) has established itself as the standard. Some manufacturers are already producing 8-in. (approx. 20 cm) solar cells. Large solar cells require fewer processing steps to be made into modules. However, there is greater risk that the cells will break during further processing. The current increases with the size of a solar cell, whereas the voltage remains constant. The electric voltage of a solar cell is only 0.6–0.7 V.

Clearly, higher voltages are needed for practical applications. Therefore, many cells are interconnected in series to form solar modules. Soldered wires are used to connect the front contacts of a cell to the back contacts of the next cell. It takes 32–40 cells connected in series to produce a voltage high enough to charge 12-V batteries. Higher voltages are needed to feed into the grid through inverters. Solar modules with at least 60 cells connected in series are common for this purpose.

Solar cells are very sensitive, break easily, and corrode when in contact with moisture, so they must be protected. Consequently, they are embedded in a layer of special plastic between the front glass panel and a plastic film on the back (Figure 5.6).

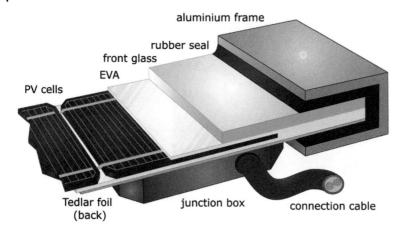

aluminium frame

rubber seal

front glass

EVA

PV cells

Tedlar foil
(back)

junction box

connection cable

Figure 5.6 Basic structure of a photovoltaic module.

Some manufacturers also use glass for the back. The glass provides mechanical stability and must be very translucent. The plastic material used for embedding the solar cells consists of two thin films of ethylene-vinyl acetate (EVA). At temperatures of around 100 °C these films bond with the cells and the glass. This process is called lamination. The finished laminate then protects the cells from the effects of the weather, especially moisture.

The connections of the solar cells are linked to a module junction box. Individual faulty cells or uneven shade can damage a PV module. Bypass diodes are designed to prevent any damage by compensating for any faulty cells. These diodes are also usually integrated into the module junction boxes.

5.2.3 Thin-Film Solar Cells

Crystalline solar cells are practically carved from solid blocks, a process which requires a comparatively large amount of costly semiconductor material. Alternative production methods using thin-film cells seek to reduce the amount of material needed. Whereas crystalline solar cells reach thicknesses in the order of tenths of millimetres, the thickness of thin-film solar cells is in the thousandths of a millimetre range. The production principle is similar even when different materials such as amorphous silicon (a-Si), cadmium telluride (CdTe), or copper indium diselenide (CIS) are used (Figure 5.7).

The base of thin film solar cells is a substrate that is usually made of glass. Plastic can be used instead of glass for the substrate to produce modules that are flexible and bendable. A thin TCO (Transparent Conductive Oxide) layer is applied to the substrate using a spraying technique. A laser or micro-cutter then separates this layer into strips. The individual strips constitute the single cells that form the basis of the solar module. Like crystalline cells, these cells are also bonded in such a way that they are connected in series to increase the electric voltage. The long strips make it visually easy to distinguish thin-film modules from crystalline solar modules.

The semiconductor and doping materials are then vaporized at high temperatures. When silicon is vaporized as a semiconductor material, its crystalline structure is lost.

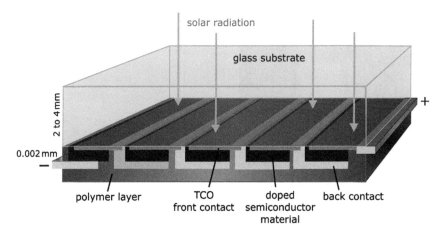

Figure 5.7   Cross-section of a thin-film PV module.

This is referred to as amorphous silicon. A screen-printing procedure then applies materials like aluminium to the contact at the back. A layer of polymer seals the cell at the back to protect it from moisture.

The efficiency of thin-film modules is currently still lower than that of crystalline PV modules. This means that a larger surface is required for the same power output, and, therefore, more assembly is required, and the associated costs are higher. However, the efficiency of individual thin-film technologies has increased significantly in recent years. Nevertheless, thin-film technology has not succeeded in breaking the dominance of crystalline solar modules.

In addition to thin-film materials, other technologies are currently being tested. Pigment cells and organic solar cells could eventually offer a cost-effective alternative to present-day technologies. At the moment it is almost impossible to predict which technologies will prove to be the best in 30 or 40 years' time. The fact is that competition stimulates business, and costs will continue to fall due to the rivalry between different PV technologies.

5.3   PV Systems – Grids and Islands

5.3.1   Sun Islands

With PV systems a distinction is made between stand-alone (island) systems and grid-coupled systems. Stand-alone solar systems work autonomously without being connected to an electricity grid. For example, they are often used in small applications, such as wristwatches and pocket calculators. This is because, in the long run, they are less expensive than using disposable batteries to supply energy, and a network cable in this case would be highly impractical. Stand-alone solar systems are also popular for use in small systems like car park ticket machines. In this case it is less expensive to install a PV system than to lay network cables and install a metre (Figure 5.8).

However, the big market for solar stand-alone systems is in areas that are far from an electricity grid. Globally, around two billion people have no access to electricity. Even in

Figure 5.8 Stand-alone PV systems offer advantages for many applications compared to grid connections.

industrialized countries there are towns that are very remote from the grid and where any kind of cabling would be extremely expensive. There are alternatives to using stand-alone solar systems, such as diesel generators. However, nowadays a solar system often compares favourably in terms of cost and supply reliability, especially in cases with low electricity demand. Nevertheless, the capital costs of stand-alone PV systems are relatively high. Diesel generators are cheaper to buy. However, due to the high cost of diesel fuels, they are usually more expensive than the solar alternative over their entire service life. In developing countries in particular, special financing models such as microcredits can help to overcome the hurdle of high capital costs and thus promote the further spread of PV.

Solar island systems are comparatively straightforward and can even be installed by people with limited technical skills (Figure 5.9). A battery ensures that supply is available at night or during periods of bad weather. For reasons of cost, lead batteries are normally used. In principle, 12-V car batteries are also an option. Special-purpose solar batteries have a considerably longer lifetime, but are also more expensive. Since batteries can quickly be ruined as a result of leakage or overcharging, a charge regulator protects the battery. The battery, the power consumer, and the PV module are connected directly to the charge controller. Take care not to mix up the positive and negative poles; otherwise a short circuit may occur.

When the battery is nearly empty, the consumer load is switched off. Although the loss of power is annoying, this is better than a defective battery. Once the battery reaches a

photovoltaic modules consumers

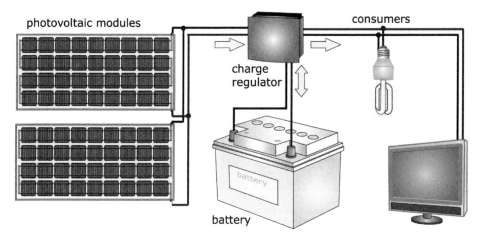

charge regulator

battery

Figure 5.9 Principle of a stand-alone PV system.

certain charge level again, the charge controller automatically switches the consumer load back on. Once the battery is fully charged, the charge controller separates the PV module and prevents the battery from overloading.

The costs can be kept low if the consumer loads are energy-efficient. A battery voltage higher than 12 V is recommended for consumers with higher loads, as otherwise the losses in the lines will be too high. Island systems work on a direct-current (DC) voltage basis, so, if possible, only DC consumer loads should be used. Special 12 or 24 V DC refrigerators, lamps, and entertainment devices are available. To use alternating current (AC) consumers, a stand-alone inverter must first convert the DC voltage of the battery to AC voltage.

Normal low voltages are relatively safe, at least as far as contact is concerned. Since batteries are power packs, improper handling can cause short circuits, fire, or even explosions. Battery rooms should always have good ventilation as hydrogen gas can build up in them. Lead batteries contain diluted acid. Over time, water evaporates from the battery, and the battery must be refilled regularly. With maintenance-free batteries the water is bound in a gel and cannot escape.

A large array of PV modules can end up costing the same as a mid-range car, and even individual PV modules are quite expensive, so thieves are making life increasingly difficult for the operators of remote PV systems. Solar modules installed on quiet roads are particularly vulnerable to theft. Since stand-alone systems are usually installed in less busy locations, the risk of theft is especially high. This risk should be minimized during installation. Ideally, PV systems should not be visible from public roads. However, if this cannot be avoided, systems should at least be installed in places that are difficult to access (Figure 5.10).

In Germany, small solar systems can only make a small contribution to climate protection. Nevertheless, PV-based pocket calculators and watches help reduce the mountain of small spent batteries. In countries with poorly developed electricity grids, however, stand-alone PV systems are becoming increasingly attractive as an alternative to diesel generators, which have dominated until now, and can play an important role in the climate-friendly development of electricity supply systems.

Figure 5.10 Typical locations for stand-alone PV systems. Left: Electricity supply for a village in Uganda. Right: Alpine lodge. Source: SMA Technologie AG.

5.3.2 Sun in the Grid

PV systems that feed into the public grid are built in a different way from island systems. First of all, they normally require a larger number of modules. Crystalline solar modules installed over a $30\,m^2$ area can yield a peak output of 4–5 kWp. In Germany this generates between 3500 and 5000 kWh with the right kind of roof; in Southern Europe or good sites in the USA the figure increases by around 50%. This can easily cover the electricity demand of an average household. An ideal surface can usually be found on the roofs of single-family houses.

Since solar modules deliver DC voltage, but the public grid works with AC voltage, an inverter is required. An inverter converts the DC voltage of a PV module to AC voltage. Modern inverters have high demands placed on them. They are required to have a high level of efficiency to ensure that, any loss of valuable solar energy resulting from conversion into AC voltage is minimal. Modern PV inverters achieve efficiencies of up to 98%. It is important that the efficiency is also high under partial load, i.e. cloudy skies. European efficiency describes average inverter efficiency based on Central European climate conditions, while CEC efficiency is for Californian conditions (Figure 5.11).

Inverters also have to monitor the grid constantly and switch off the solar feed when a general grid outage occurs. Otherwise, if the electricity supply company wants to carry out work on the grids and switch off the power for this purpose, the solar power fed into the grid could endanger the workers. If a public power outage occurs, owners of PV systems will unfortunately end up sitting in the dark even though there is nothing wrong with their solar system. Technically, a solar system can be kept running in island mode, using batteries, if the grid fails. This option will be described in more detail later.

An inverter does not only convert voltage. It also ensures that PV modules operate with the optimal voltage and deliver the maximum possible power. The optimal voltage setting is referred to as MPP tracking. At the system design stage, it is important to match the number of PV modules to the inverter. The leading inverter manufacturers usually offer relevant design software free of charge.

Shadowing can also cause problems. PV systems react sensitively, and performance suffers even if only part of a system is in the shade. If three cells are in shade, the entire

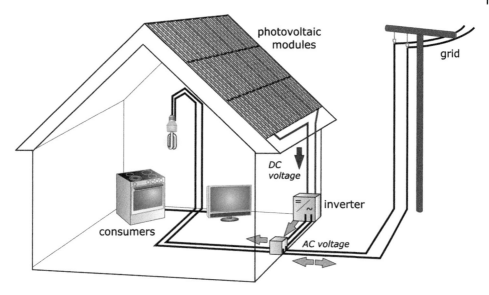

photovoltaic
modules

grid

DC
voltage

inverter

consumers

AC voltage

Figure 5.11 The principle of a grid-connected PV system.

PV module can stop working. Therefore, a site that has minimal shade is more important for the installation of a system than an optimal orientation towards the sun.

Grid-connected PV systems feed all the electricity they generate into the public grid. They act as solar power plants. Solar systems can also track the sun (Figure 5.12). Systems that use tracking can increase power output by 30% on average over the year. However, tracking also increases the capital cost and requires additional maintenance due to

Figure 5.12 The 'Gut Erlasee' tracked solar power plant in Bavaria, Germany, has a total output of 12 MW. Source: SOLON SE, Photo: paul-langrock.de.

Figure 5.13 PV façade system. Source: Photo: SunTechnics.

the mechanical parts required. Since module prices have fallen sharply in recent years, tracking systems are quite rare these days.

The neatest looking installations of PV systems are those on roofs and façades (Figures 5.13 and 5.14). Less building material is needed for this type of installation, which is a cost advantage of PV systems. And compared to a representative marble façade, PV is now available at a bargain price.

Figure 5.14 PV systems on single-family homes. Source: Photo: SunTechnics.

Some of the solar power can be used directly in the building when the system is installed. If the PV system produces more power than the building needs, it feeds the surplus electricity into the public grid. If the output of the solar system is not sufficient to cover the building's own requirements, the electricity shortfall is taken from the grid. In a sense, the grid acts as a storage unit. Strictly speaking, however, the grid cannot store power. When solar power is fed into a grid, then other power plants cut their production. As a result, solar systems reduce the emissions of existing power plants. If there is insufficient output, this then has to be sourced from other power plants. These power plants do not necessarily have to be coal-fired, gas, or nuclear plants. On the contrary, PV systems work compatibly with other renewable energy plants, such as wind power, hydropower, or biomass plants.

Connecting to the grid in Germany is usually not a problem. An electrician is needed to connect the system, and the relevant electricity supplier has to be notified. In addition to the connection protocol, technical documents relating to the PV system are submitted with the application. It is important that the system complies with the general regulations, and this is usually the case with common system suppliers. Usually a representative of the electricity supply company then inspects the system. Payments are handled automatically and are calculated based on the metre reading. The electricity company and the operator of the PV system usually also sign a contract. However, this is not always mandatory.

A separate electricity metre is required to calculate the output from the PV system. This metre calculates the quantity of electrical energy that has been fed into the grid, which is then credited according to existing tariffs. Since the remuneration for solar power in Germany for new systems is now below the electricity prices for households and smaller commercial enterprises, it makes sense to consume as much solar power as possible directly (referred to as 'self-consumption') and to minimize the surplus fed into the grid.

5.3.3   More Solar Independence

With a self-consumption level of 100%, no solar power would be fed into the grid at all. In practice, the achievable self-consumption level is usually considerably lower. However, it can be increased in a targeted manner by running large consumer loads such as the washing machine around midday, when the solar radiation is at its maximum. Various manufacturers now also offer devices for automatic consumption control. However, even with such measures, the options for increasing the self-consumption level are limited.

> ### Self-Consumption and Self-Sufficiency
>
> If the solar power of a PV system is consumed directly on site, this is referred to as self-consumption. The self-consumption level indicates how much of the solar power is consumed directly and not fed into the grid. Self-consumption reduces the need to buy expensive mains electricity. Since the prices for buying electricity are usually significantly higher than the payments for feeding solar power into the grid, the profitability of a PV system increases with increasing self-consumption. However, very high self-consumption

levels can generally only be achieved with very small PV systems or additional storage facilities.

The degree of self-sufficiency indicates what proportion of one's own electricity demand is covered by a PV system. With 100% self-sufficiency, electricity is no longer taken from the grid and you are completely independent of energy suppliers and any variation in electricity prices. However, complete self-sufficiency is virtually impossible to achieve in Germany at a justifiable cost. For high degrees of self-sufficiency, large PV systems and large storage facilities are generally required in order to make a large proportion of solar energy available at night and in winter. A solar system with such a configuration produces large surpluses during the day in summer, which have to be fed back into the grid, which in turn reduces the self-consumption level.

When planning a solar system, a compromise must therefore always be sought between the desire for a high degree of independence with a high degree of self-sufficiency and good economic efficiency with a large self-consumption level.

Very high self-consumption levels can generally only be achieved with very small PV systems. However, these can then only cover a very small part of their own electricity demand and thus only achieve very low degrees of self-sufficiency. This means that little solar power is fed into the grid and almost all the output is used during the day, while at night and on cloudy days electricity demand from the grid remains high.

If a storage facility is combined with the PV system, significantly larger systems with higher degrees of self-sufficiency can be configured, which also benefit from high self-consumption levels. A wide range of battery systems are available for this purpose (Figure 5.15).

In battery systems, too, priority is given to using solar electricity directly on site. Any excess is used to charge a battery. Only when the battery is full does the system feed

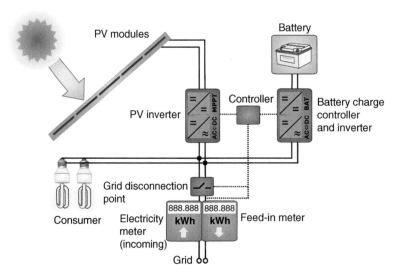

Figure 5.15 Grid-connected PV system with battery storage to increase the self-consumption level [Qua13].

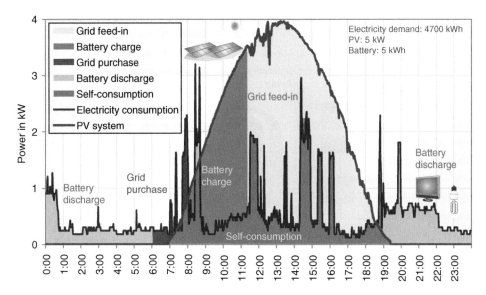

Figure 5.16 Power flows in a grid-connected PV battery system for a household in a detached house on a sunny Sunday in spring.

solar electricity into the grid. If the PV modules supply less electricity than is required on site, the battery initially covers the shortfall. Once the battery is empty, mains power is used to keep the system running (Figure 5.16). Modern battery systems can also relieve the public grid. For this purpose, they are not charged with the first surpluses from the solar system as shown in Figure 5.16. A forecast algorithm determines when the largest surpluses from the PV system will occur during the day and then stores them in the battery system instead of feeding them into the grid.

A battery system can also be designed so that it can be disconnected from the mains via a disconnection point in the event of a power failure. This means that it can continue to operate as a stand-alone system and, with the aid of the battery, ensure the power supply for a certain period of time. It then operates as an emergency power system and increases the security of the supply.

Lead-acid or lithium batteries can be used. Lead-acid batteries are familiar as starter batteries for cars. While lead-acid batteries are still frequently used in smaller stand-alone PV systems, lithium batteries have become established in grid-connected systems. Lithium batteries have a significantly longer service life. Lead-acid batteries can last for 5–10 years if used properly, lithium batteries for 20 years. They are also less sensitive to deep discharges. Battery rooms with lead-acid batteries must always be well ventilated, since the batteries can emit hydrogen, which can lead to the formation of oxyhydrogen gas.

Batteries have become standard components in many PV systems; their popularity has risen thanks to sharp cost reductions over recent years. Systems with hydrogen storage are an alternative to battery systems. Excess electricity is converted to hydrogen with the aid of an electrolysis unit and, if required, converted back to electricity via a fuel cell. Since these systems are significantly more expensive than battery systems, their market shares will remain low for the foreseeable future.

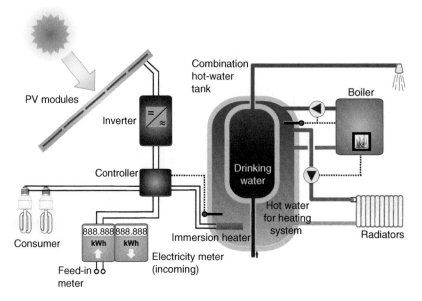

PV modules

Inverter

Controller

Consumer

888.888 888.888
kWh   kWh

Feed-in
meter

Electricity meter
(incoming)

Immersion heater

Combination
hot-water
tank

Drinking
water

Hot water
for heating
system

Boiler

Radiators

Figure 5.17 Coupling of a PV system with a conventional heating system becomes interesting when the feed-in tariff falls below the fuel costs. Source: [Qua13].

Another conceivable solution for increasing the share of self-consumption is coupling with an existing heating system (Figure 5.17). If a reasonably large heat store is available, an electric heating element can be retrofitted comparatively inexpensively. In principle, a more efficient heat pump can also be used instead of the heating element. However, since heat pumps are considerably more expensive, this significantly increases the capital costs and usually makes such a system economically unattractive, despite their greater efficiency. However, if a heat pump happens to be already available, then coupling with the PV system is essentially a 'no-brainer'.

Solar surpluses can then be used to heat the storage tank and thus save fuel in the heating system. In this way, the PV system can cover a larger part of the hot water demand and also supplement space heating in the spring and autumn. However, even with such a system, the majority of the heating demand is still covered by the conventional heating system and not by the PV system. It only makes economic sense to use the surplus solar energy when the tariff for the solar electricity fed into the grid falls below the fuel prices for heating.

5.4 Planning and Design

5.4.1 Designing Stand-Alone Systems

The design of stand-alone PV systems is fundamentally different from grid-connected systems. They cannot rely on the public electricity grid if there is no sun. Instead these systems use sufficiently large batteries to avoid any power failures. However, a battery should essentially only be used to bridge a small number of challenging days. Therefore, PV modules have to deliver as high a yield as possible during the months with the

Table 5.3 Monthly and yearly totals of solar radiation in kWh/m² for different locations and orientations

Location	Berlin		Freiburg		Málaga	Cairo
Orientation	30° south	60° south	30° south	60° south	60° south	50° south
February	50	55	64	51	146	161
April	151	143	151	138	157	183
June	170	135	165	134	156	201
August	148	134	153	135	183	196
October	73	80	90	96	177	205
December	21	33	37	44	148	189
Year	1234	1168	1329	1248	1967	2230

Radiation data: average between 1998 and 2010.
Source: PVGIS, http://re.jrc.ec.europa.eu/pvgis.

least sunshine. The recommendation for reliable operation in winter is to place PV modules at a steeper angle than is necessary with grid-connected systems, which are angled for optimum operation all year round. In Europe and North America, a tilt of around 60°–70° towards the south provides optimal solar yield in the month of December. The closer one gets to the equator, the less marked are the differences between summer and winter. In these parts of the world a shallower installation is sufficient for winter operation. In Cairo, for example, the optimum angle is around 50°. In Nairobi, however, an almost horizontal installation is recommended (Table 5.3).

The aim of a stand-alone system is not to achieve the highest possible yield with a solar system, but to supply certain consumer loads reliably. Therefore, the system design should be based on the month with the poorest supply. In Germany, this is December. Furthermore, a solar system should have a safety margin of at least 50%. As with grid-connected systems, the performance ratio (*PR*) takes losses into account.

 Required PV Module Output for Stand-Alone Systems

The required MPP output P_{MPP} of the PV modules can be calculated approximately from the solar irradiation $H_{solar,m}$ in the worst month in kWh/m², the electricity demand $E_{demand,m}$ in the same month, a safety margin f_S of at least 50% and the performance ratio PR (0.7 on average):

$$P_{MPP} = \frac{(1+f_s) \cdot E_{demand,m}}{PR} \cdot \frac{1\frac{kW}{m^2}}{H_{solar,m}}.$$

The battery should be dimensioned in such way that it is usually only half discharged and can meet the total electricity demand for a number of 'reserve days'. In Central Europe up to five reserve days dR are enough to ensure reliable operation in the winter; in countries that get more sun two to three reserve days would be sufficient. If snow-covered PV modules are expected to be unable to supply electricity for a longer period of time in the winter, even more reserve days will be necessary. The required battery capacity is calculated on the basis of the battery voltage V_{bat} (e.g. 12 V):

$$C = \frac{2 \cdot E_{demand,m}}{U_{bat}} \cdot \frac{d_R}{31}$$

For example, a PV system should be capable of operating an 11 W low-energy light bulb in a summerhouse for three hours every day in the winter. The monthly electricity demand for one month is then $E_{demand,m} = 31 \times 11 \text{ W} \times 3 \text{ h} = 1023 \text{ Wh}$. With a safety margin of 50% = 0.5 and a performance ratio of 0.7, a module in Berlin orientated towards the south and tilted 60° then has a required MPP power of

$$P_{MPP} = \frac{(1+0.5) \cdot 1023 \text{ Wh}}{0.7} \cdot \frac{1\frac{\text{kW}}{\text{m}^2}}{32\frac{\text{kWh}}{\text{m}^2}} = 68.5 \text{ W}.$$

With five reserve days and a battery voltage of 12 V, the battery capacity is

$$C = \frac{2 \cdot 1023 \text{ Wh}}{12 \text{ V}} \cdot \frac{5}{31} = 27.5 \text{ Ah}.$$

5.4.2 Designing Grid-Connected Systems

The first thing to do when planning a solar system is to check whether a system can actually be installed. From the perspective of planning and building laws and regulations, solar systems are structural components – even if they are simply screwed to the roof of a house. The authority responsible for local building regulations determines whether approval is necessary, and, if so, what type of approval. In most cases PV systems do not need approval from the local authority unless they are erected on a greenfield site. Complications arise when the sites for such systems are subject to architectural conservation laws. A permit from the relevant authority is required if a solar system is to be installed on or near a protected building. In addition to obtaining planning permission, it is always a good idea to consult the building regulations and the development plans of the local authority. These plans will include any conditions that the local authority has stipulated for the construction of PV systems.

If there are no legal obstacles, the planning phase can start. If a PV system is to be installed on the roof of a house, you must first determine which parts of the roof can be used for a solar thermal system (see Chapter 6). As PV systems are sensitive to shade, it is recommended that the part of the roof used to generate solar power is shade-free. The installation should not be placed near chimneys, aerials, and other roof structures.

Available PV Capacity

Once the type of solar module has been selected, an approximate calculation of the available PV capacity can be made, based on the remaining roof area A and the efficiency (see Table 5.1):

$$P_{MPP} = A \cdot \eta \cdot 1\frac{\text{kW}}{\text{m}^2}.$$

The available capacity on a useable area of 27.8 m² with a module efficiency of 18% (0.18) is

$$P_{MPP} = 27.8 \text{ m}^2 \cdot 0.18 \cdot 1\frac{\text{kW}}{\text{m}^2} = 5 \text{ kW}_p.$$

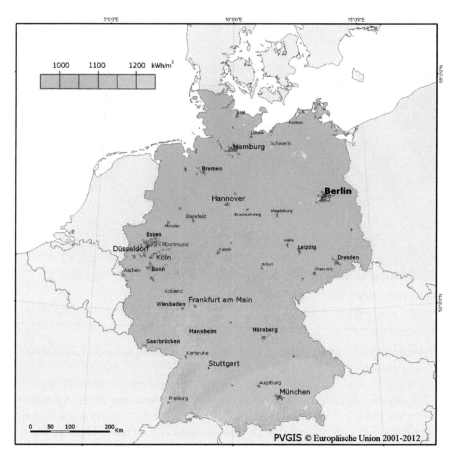

Figure 5.18 Mean annual total solar radiation energy in Germany in kWh/m² between 1998 and 2011. Source: PVGIS, http://re.jrc.ec.europa.eu/pvgis [Sur07, Hul12].

The annual system yield can be calculated based on this capacity. We must first determine the available solar energy. The map in Figure 5.18 shows the annual total solar radiation in Europe based on averages over many years. The radiation data for the USA can be found at http://www.nrel.gov/rredc. Fluctuations of over 10% in solar availability are possible between the individual years given. Since the 1980s, annual irradiation in Germany has increased by 5–10% due to a decrease in air pollution. Therefore, only relatively new radiation data should be used for the system design.

However, the values only apply to the horizontal orientation of a system. If a system is mounted on a sloping roof, the roof determines the orientation of the system. In Europe and North America, the roof should ideally be tilted around 35° to the south. With an optimal orientation towards the sun the supply of solar radiation increases by around 10%. However, good radiation values are still achievable even if the orientation is less favourable. Figure 5.19 shows the tilt gains and losses for all possible orientations for Berlin. These values can also be applied to other locations in central Europe.

If a PV system is to be installed on a roof or a greenfield site, the PV modules can be orientated optimally 35° towards the south. If the solar modules are set up in several

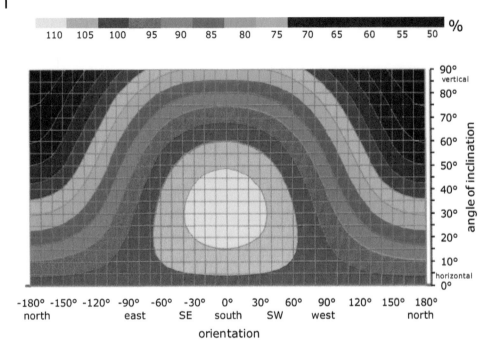

Figure 5.19 Change in annual solar radiation in Berlin depending on orientation and tilt angle of PV systems.

rows one behind the other, they will shade each other as the sun goes down. A distance at least twice the module height should therefore be maintained between the module rows. This means that only around one-third of the area can be used. The loss due to shade then usually amounts to less than 5%.

The proportion of solar radiation that PV modules convert into electric energy depends on the quality of the system. Solar modules rarely achieve the rated power indicated. Dust, bird droppings, increases in air temperature, line losses, reflection, inverter losses, and other factors reduce performance. The relationship between real efficiency and rated efficiency is called the performance ratio (*PR*). Table 5.4 lists the criteria for performance ratio values for grid-connected PV systems.

Table 5.4 Performance ratio of grid-connected PV systems

Performance ratio *PR*	Description
0.85	Top-quality system, very good ventilation, no shade, minimal pollution
0.8	Very good system, good ventilation, no shade
0.75	Average system
0.7	Average system with some losses due to shade or poor ventilation
0.6	Poor system with significant losses due to shade, pollution, or system failures
0.5	Very poor system with large areas of shade or defects

Electrical Energy Yield of Grid-Connected PV Systems

A rough calculation of the quantity of energy fed in annually by a grid-connected PV system can be made on the basis of the annual solar radiation H_{solar} in kWh/(m² a) from Figure 5.18, the losses or gains due to orientation and tilt angle f_{tilt} (Figure 5.19), the rated or MPP power P_{MPP} of the PV modules in kWp and the performance ratio PR (cf. Table 5.4):

$$E_{electrical} = \frac{H_{solar} \cdot f_{tilt} \cdot P_{MPP} \cdot PR}{1\frac{kW}{m^2}}.$$

The yield of an unshaded PV system in Berlin with a tilt of 20° and an MPP output of 5 kW$_p$ is calculated here as an example. The annual solar radiation according to Figure 5.18 is around 1075 kWh (m² a)$^{-1}$. According to Figure 5.19, the tilt gains are $f_{tilt} = 110\% = 1.1$ based on a 20° south–south-west orientation. Based on the performance ratio of an average system $PR = 0.8$, the annual solar electricity yield is:

$$E_{electrical} = \frac{1075\frac{kWh}{m^2 a} \cdot 1.1 \cdot 5 kW_p \cdot 0.8}{1\frac{kW}{m^2}} = 4730\frac{kWh}{a}.$$

This corresponds to approximately the consumption of an average German single-family house. Therefore, as a yearly average about 30 m² of roof area is needed for a household to cover its total electricity requirements using a solar system. The specific yield related to 1 kW$_p$ is often given for a system. In the example above, the total is 946 kWh (kW$_p$ a)$^{-1}$. ■

The rough calculation of yield described here naturally incorporates certain inaccuracies. However, the order of magnitude of the system yield is correct. It can be used in the following section to examine the economic viability. Internet tools and computer programs are available to conduct a more precise analysis *(see web addresses below)*. Professional firms also offer to carry out calculations of this type. In any case, if a large system is to be installed, it is advisable to get an expert to produce an appraisal on yield to avoid unpleasant surprises later on. Banks also often require a copy of the relevant appraisal.

- valentin.de/calculation/pvonline/pv_system   Online calculations for PV
- re.jrc.ec.europa.eu/pvgis   systems
- www.polysunonline.com

It is not necessary for a grid-connected system to be large enough to cover the total electricity demand of a household. In the past, the size of the PV system was usually determined by the roof area. Quite often, the system capacity significantly exceeded the self-consumption level. In the future, systems that are optimized for self-consumption will become the norm. Large systems with a high degree of self-sufficiency not only promise increased independence from the energy supplier and its constantly rising electricity prices; but also make an important contribution to climate protection when they are installed in their millions.

5.4.3   Planned Autonomy

In the past, no further calculations were undertaken at this point, since all the solar power was fed into the grid and remunerated. Further considerations were therefore unnecessary. Today, however, the aim is to achieve high self-consumption levels to ensure good economic efficiency. In other situations, the aim is to achieve high degrees of self-sufficiency. This means one first needs to determine which proportion of the solar electricity generated is consumed locally and which is fed into the grid. The self-consumption and self-sufficiency levels depend on various parameters. The main influencing factors are the amount of electricity consumed and the size of the PV system. If a battery is added, the battery capacity naturally also plays a role.

The electricity bill makes it easy to determine one's own electricity consumption. The time period when the electricity is consumed also plays a role. In a pensioner's household, where electric cooking takes place every day at noon, the self-consumption level is higher than in the case of two working people who are mainly at home in the evening and at weekends. Compared to the other factors already mentioned, however, the type of household plays a lesser role. Therefore, to keep things simple, this influence is not considered.

The self-consumption and self-sufficiency levels can be determined quite easily from Figure 5.20. For example, if a PV system with an output of $5\,kW_p$ is to be installed on a single-family house with an annual electricity consumption of 5000 kWh, the ratio of PV capacity in kW_p to electricity consumption in 1000 kWh is 1.0. If no battery is provided, the usable battery capacity is zero. The self-consumption level is 30%. In other words, 70% of the solar electricity generated must be fed into the grid as excess. The self-sufficiency level is also 30%. This means that 70% of the local electricity consumption is still covered by an energy supplier via the grid. If a battery with a usable capacity of 5 kWh is added, the ratio of battery capacity to electricity consumption is also 1.0. This increases the self-consumption level to around 60% and the self-sufficiency level to over 50%.

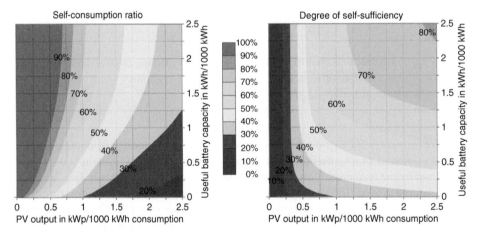

Figure 5.20 Achievable self-consumption and self-sufficiency levels of PV self-consumption systems in single-family houses with battery storage [Wen13].

The self-consumption value determined in this way can then be used for the subsequent economic considerations. The self-consumption level reduces the consumption of electricity from the grid and reduces the electricity bill accordingly. Surpluses are fed into the grid based on the corresponding tariff. The cost-effectiveness of the PV system is based on mixed financing from saved electricity costs and feed-in revenues.

From Thinking About Having a PV System on the Roof to Getting a Quote

- Determine the orientation and tilt angle of the roof. Is the roof suitable in terms of orientation and tilt angle? Recommendation: at least 95% according to Figure 5.19 on page
- Establish shading conditions on the roof. Does my roof have any shade? Recommendation: take shaded areas into account during planning; also pay attention to chimneys, aerials, and lightning conductors. Identify available roof area with no shade.
- Calculate installable MPP power for the PV system. Recommendation: Efficiency at least 16% in formula on page
- Calculate possible annual electric energy yield. Use the formula on page or online tools.
- Is the idea to increase the security of supply and the degree of self-sufficiency by using a battery?
- Determine the self-consumption level according to Figure 5.20 on page.
- Request quotes.

5.5 Economics

In some niche applications, PV has long been fully economically competitive. Small applications are often supplied with small batteries or button cells. Compared to household electricity prices of around 30 cents kWh^{-1} in Germany, for example, with PV the costs can quickly explode to several hundred euros per kWh. It takes around 280 AA-size high-quality alkaline manganese battery cells to store 1 kWh. However, no-one would ever consider buying 280 AA-size batteries to run a washing machine just once. Yet with small applications we often tend to be willing to pay whatever it costs to buy batteries. In many cases it is the infrequent use of these small applications that makes using electricity affordable in the first place. PV can compete with such high energy cost even under the most adverse conditions. Even with larger battery systems, PV is often an economical alternative. However, in order to effectively protect the climate, PV must do more than just domestic applications. This is only possible with grid-connected systems that replace conventional power plants.

5.5.1 What Does It Cost?

The minimum size for a grid-connected PV system is a PV module with about 0.3 kW$_\text{p}$ output. There is no upper limit; this depends only on the area and the amount of money available. Fortunately, the prices of PV systems are dropping so much that they are out of date even before they are printed. This makes it difficult for any published work to provide a current price guide. In 2005 the capital cost for a fully installed 5 kW$_\text{p}$ system without battery storage was around €30 000. In 2018, the figure was just €7000 excluding sales tax. The price of the PV module itself only constitutes part of the total system price.

Around 50% of the capital cost is accounted for by the PV modules, while the rest is spent on inverters, mounting materials, the actual mounting, and planning. Prices for batteries have also fallen significantly in recent years. Further cost reductions are expected in the coming years.

Whereas the lion's share of the PV system cost is in the installation, the revenues then come through the electricity that is generated. The payment is usually per kWh. The operating costs of PV systems are comparatively low. Each year an estimated 2–5% of the capital cost is spent on running costs such as insurance, possible leasing, metre rental, and reserves for repairs. The modules can normally be used for 20–30 years. Repairs or replacements should be taken into account right from the start.

Another factor that has a major effect on generation costs is the return. Only very few idealists will invest their own capital in a PV system in the hope of at best receiving their invested capital back again over the lifetime of the system. The investment should yield a little more than a savings account in order to become attractive for larger groups of people. Even for this you need a certain amount of idealism, because the risk of a PV system is usually regarded as higher than that of a savings book. On the other hand, savings accounts are no longer what they used to be. Bank failures resulting in total loss of savings can no longer be ruled out.

A PV system, on the other hand, can be destroyed by a lightning strike or a storm. If the insurance does not pay for the damage, the investment is lost. Also, the yield may be less than predicted. There may be various reasons for this. The PV system could be the victim of poor planning, over the years a tree could grow so that it shuts out the sun, a flock of birds could regularly be using the module for target practice, or the solar radiation may be lower than during the previous 20 years due to volcanic eruptions. Ultimately, it is the operator of the PV system who is responsible for dealing with all these risks. A return of around 6% is therefore reasonable for an investment that is not driven by idealism. Conversely, self-consumption PV systems help to keep one's own electricity costs stable. The higher the price rises imposed by electricity suppliers, the more economical the PV system becomes.

Figure 5.21 shows the resulting generating costs of a PV system without battery storage, depending on the net capital costs. The assumption made in the calculations is that 3.5% of the investment costs is for operating and maintenance costs each year and the economic lifetime of the system is 20 years. The different coloured lines represent the calculations for various specific yields. In Germany these are usually less than $1000 \, \text{kWh} \, \text{kW}_p^{-1}$. For a return of 6% and net investment costs of €1500 per kW with a specific yield of $1000 \, \text{kWh} \, \text{kW}_p^{-1}$, the resulting generation costs are $18.3 \, \text{cents} \, \text{kWh}^{-1}$. If the expected return is zero, $12.8 \, \text{cents} \, \text{kWh}^{-1}$ is sufficient. These costs must be generated on average from the electricity costs saved through self-consumption and the feed-in tariff.

5.5.2 Funding Programmes

While smaller stand-alone PV systems have long been competitive without any subsidies, larger grid-connected PV systems usually only pay for themselves if feed-in tariffs are available. Smaller systems can pay for themselves through the saved electricity costs, but here too an appropriate remuneration for feeding in surpluses is helpful. In

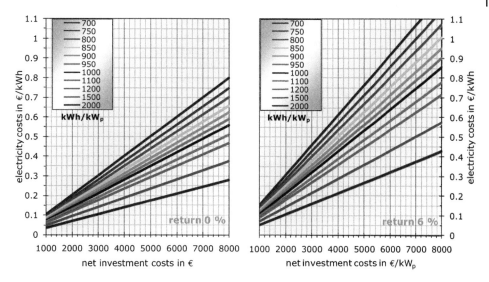

Figure 5.21 Electricity generation costs as a function of net capital costs and specific yield for a return of 0% (left) and 6% (right).

Germany, the remuneration is determined by the Renewable Energy Sources Act (EEG). It prescribes a fixed price for every kilowatt hour that is fed into the public electricity supply grid by PV systems. The remuneration must be paid by the electricity supply company responsible, who may in turn pass on the additional costs incurred to all electricity customers. One aim is to make solar power competitive. Therefore, the law provides for a steady degression. The remuneration for new installations therefore decreases continuously. The remuneration is granted for a period of 20 years from the installation of the plant. The amount of remuneration decreases with the size of the system. Since 2016, PV systems with an output of more than 750 kW have had to successfully submit a tender application in order to receive funding.

In 2012, grid parity was achieved. Since then, PV electricity has been cheaper than domestic electricity from the grid (Figure 5.22). This has fundamental effects on the PV market. While large PV systems were previously planned exclusively for grid feed-in and were not economical without feed-in tariffs, they are increasingly profitable due to electricity savings in self-consumption. In the meantime, even the difference between the remuneration for PV systems and the price of heating oil has become relatively small. In the foreseeable future, the feed-in tariff for solar power is expected to fall below the fuel costs for heating oil. This will certainly be the case for all PV systems for which the right to EEG compensation expires after 20 years. PV will then also become attractive as a supplement for heating systems.

The increased remuneration under the Renewable Energy Sources Act in Germany is expected to be phased out in the longer term. However, in order for PV to be completely competitive without subsidies, storage systems must become significantly cheaper. In order to achieve this, low-interest loans are envisaged for storage introduction programmes. In the past, low-interest loans were also available for the construction of PV systems in addition to the EEG remuneration, which are handled by the German Kreditanstalt für Wiederaufbau (KfW). Since the subsidy programmes are subject to rapid

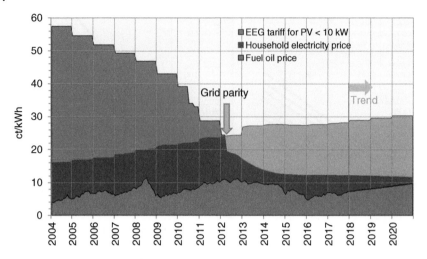

Figure 5.22 Development of the EEG remuneration in Germany for small PV systems with outputs below 10 kW compared to the household electricity price and the fuel costs for oil heaters from 2004 to 2017 and trend from 2018 onwards.

changes, it is advisable to enquire about the current conditions before installing a system *(see web tip)*.

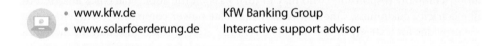

- www.kfw.de — KfW Banking Group
- www.solarfoerderung.de — Interactive support advisor

5.5.3 Avoiding VAT

In Germany, a PV system for feeding into the grid can be operated privately or commercially. Losses can be written off from tax via a profit and loss account; profits must be taxed as income. This applies to both private and commercial operation. The purchase of the PV system may not be fully written off as a loss in the first year of operation, but it must have been fully paid off over 20 years.

The commercial operation of a PV system is particularly interesting in view of the Value Added Tax (VAT) refund. In principle, income from large PV systems is subject to VAT. For smaller plants with annual sales of less than €17 500, section 19 of the Value Added Tax Act applies, according to which the operator does not have to pay VAT and in return waives the reimbursement option. However, everyone can VAT-register, even if the threshold is not reached. In this case, the operation of the PV system must be reported to the tax office. As a rule, the tax office also expects proof of the intention to make a profit. Once you are VAT-registered, you have to add VAT on all goods and services sold, and then pay it to the tax office. However, the VAT is refunded for purchases. In the case of PV, VAT can then be refunded for the purchase of the system or repairs. This can quickly amount to a few hundred or thousand euros. However, the goods sold, i.e. the solar electricity, will then also be subject to VAT in the same way as the self-consumed solar electricity. VAT is added to the tariff set by the Renewable Energy Sources Act and thus does not reduce revenues.

However, the VAT reimbursement involves some bureaucratic effort. A VAT return is then due annually. For large plants with annual turnover tax revenue of more than €512 or with a combination of a small plant and a complicated tax official, a turnover tax return has to be submitted every month in advance. A wide range of computer programs are available to create tax returns and submit them electronically to the tax office.

Incidentally, there is no requirement to register small PV systems as a business, despite the voluntary VAT liability. This would, in any case, makes no sense, since business tax is only levied for profits over €24 500.

 From the Quotation for a PV System for Grid Feed-in to the Installed System

The applicant should:

- determine public subsidies and low-interest loans *(see web tip)*;
- check cost-effectiveness on the basis of the offer (see Figure 5.21);
- register the planned installation with the energy supplier;
- arrange for the system to be installed by a qualified company;
- send the report completed by the electrician and technical system documents to the responsible energy supply company (ESC). ESC approves the plant and offers a feed-in contract in accordance with the EEG.
- report the system to the Federal Network Agency; (app.bundesnetzagentur.de/pv-meldeportal)
- check the feed-in contract, amend it if necessary and sign it;
- report the commercial operation of the PV system to the tax office;
- have VAT paid via VAT return reimbursed as pre-tax deduction;
- read the metre readings at the intervals agreed with the RU and pass them on to the ESC. Create invoice if appropriate;
- if necessary, send monthly VAT return to the tax office; and
- send annual profit and loss statement with income tax and VAT return to the tax office.

5.6 Ecology

There is a persistent rumour that it requires more energy to be used for the production of the silicon needed to make the solar cell than the cell itself can generate during its entire service life. Where this rumour came from is a mystery. It was probably invented by opponents of solar technology at a time when there were no detailed studies on the issue available.

It is true that the production of solar cells is relatively energy-intensive. Temperatures of well over 1000 °C are required for the production of silicon and subsequent purification. In contrast to conventional coal, gas, or nuclear power plants, however, no further energy input is required for the operation of PV systems. After commissioning, the solar system starts to return the energy required for its production. Various scientific studies in the 1990s have shown that in Germany it takes around two to three years for solar cells to generate the energy that was needed to produce them [Qua12], significantly less than two years in southern Europe. In the interim, thanks to the increased efficiencies and the reduced cell thicknesses, this time is likely to have decreased even further. With a service life of 20–30 years, a solar cell generates many times the energy that was required for its production.

Today, a 0.3 mm thick, 6-in. crystalline silicon solar cell weighs around 16 g. The energy consumption for the production of solar cells will decrease continuously, as attempts are made to drastically reduce the use of materials and thus also the costs. For example, in thin-film cells, the use of materials is already significantly lower. In the future, it can be expected that solar systems will recoup the energy needed to produce them within a few months.

Various chemicals, some of which are toxic, are used in the manufacture of solar cells, which is why, as with all chemical plants, strict care must be taken during production to ensure that no chemicals escape into the environment. However, in modern production plants with high environmental standards, this should be possible without any problems.

The finished solar system, on the other hand, is much less problematic. Nevertheless, it would be a pity to simply scrap old solar systems, as they contain valuable raw materials. The solar industry is therefore continuously improving its processes for recovering materials. Solar modules are broken down into their components, and the materials are separated by type.

5.7 PV Markets

PV offers the most versatile application options of all the renewables. Their big advantage is the modular structure. Almost all desired generator sizes can be realized, starting from the milliwatt range for the power supply of pocket calculators and watches, up to the gigawatt range for the public power supply. While pocket calculators supplied with PV energy became widespread decades ago, large-scale systems that feed solar power into the public grid only became popular with the successful market launch programs in Japan and Germany. Government schemes in both countries boosted the production of PV modules and have contributed towards annual market growth rates between 20% and 80% since the early 1990s (Figure 5.23). In the meantime, other countries have also created attractive conditions for the construction of grid-connected PV systems. While in 1980 85% of solar modules were still manufactured in the USA, this proportion had shrunk to below 10% by 2005. Japan and Germany have lost the rank of the leading PV nations and were replaced by China from 2010. Today, China dominates the global PV market.

In Germany, the spread of grid-connected PV systems began in the early 1990s with the so-called 1000 Roofs Programme. With the help of state schemes more than 2250 PV systems were erected, mainly by private households. When the programme was phased out, the use of PV began to stagnate. It was not until 2000 that fresh impetus was given through the introduction of the Renewable Energy Sources Act (EEG). Until 2012, the German PV market was the world's leading market and German solar companies were the world leaders.

In 2013, the German government significantly worsened the conditions for PV systems. The result was a radical market slump from 7.6 GW in 2012 to just 1.5 GW in 2016,

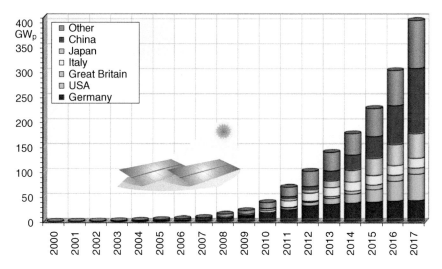

Figure 5.23 Development of total PV capacity installed worldwide.

resulting in a large number of German solar companies filing for insolvency. Around 80 000 jobs were lost in Germany.

At the end of 2017, well over one million PV systems, with an output of over 43 GW, were in operation in Germany. These systems fed 40 billion kilowatt hours of electric power into the grid, which was a 7% share of the overall electricity supply.

While from 2013 onwards Germany virtually put thumbscrews on the PV market, the PV expansion in China boomed at a breathtaking pace during the same period. In 2017 alone, China erected more PV systems than Germany had erected in the previous 30 years combined. China is unlikely to relinquish its dominant position in the coming years and is likely to remain the pacesetter in the field of PV.

5.8   Outlook and Development Potential

While just a few years ago PV was by far the most expensive way of generating electricity, today new PV systems can already outperform conventional power plants in many regions of the world. If the price decline continues, PV could become the most important type of electricity generation in the next 10–20 years.

Past experience has shown that major cost reductions are possible. Whereas the price of PV modules was still around 60 (inflation-adjusted) US dollars per watt in 1976, by 2012 it had already dropped to around $1\,W^{-1}$ and by 2017 to 50 cents. This development over the years can be seen in a so-called learning curve (Figure 5.24). What is crucial for cutting costs is an increase in production. If production quantities rise, then costs will drop noticeably, due to the effects of streamlined production and also because of technical advances. During the past 30 years cost savings of around 20% have been achieved due to the total quantities of PV modules produced being doubled. There is nothing to indicate that this development will not continue.

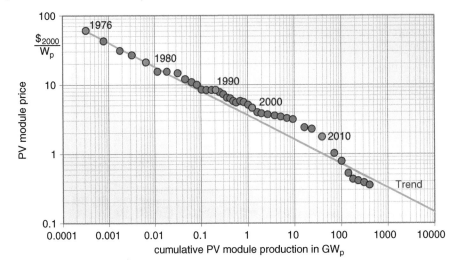

Figure 5.24 Development of inflation-adjusted photovoltaic module prices as a function of the total volume of modules produced worldwide.

In purely mathematical terms, PV could even cover the entire global energy demand. This would only take a fraction of the surface of the Sahara Desert to accomplish. Even countries like Great Britain, Germany, and France would be able to cover all their electricity requirements through PV. On the other hand, from a technical perspective, it is not a good idea to rely solely on one technology for our energy supply of the future. PV systems work well in combination with other renewable energy systems, such as wind power, hydropower, and biomass systems. A sensible combination can increase supply security and avoid the construction of large storage facilities to ensure supply at night or in winter (Figure 5.25).

Figure 5.25 De-centralized PV systems can be set up by the electricity customers, directly, who thus become competitors to conventional energy suppliers and drive forward the energy transition.

PV has the potential to become the driving force behind the global energy transition. In contrast to larger power plants, PV can also be locally installed, directly with the end customer. The electricity customers themselves can thus become direct competitors to the energy suppliers and drive forward the transformation of the energy supply system. If this development is not deliberately slowed down, PV could supply more than half of the electricity generated worldwide by the middle of the century and thus make a decisive contribution to climate protection.

6

Solar Thermal Systems – Year-Round Heating from the Sun

The light and warmth of the sun give us a special sense of well-being, which is why summer is eagerly awaited in the colder regions of the world such as Northern Europe. Most people in these countries are also lucky enough to be able to reproduce these conditions even in the depths of winter. Our homes and workplaces are heated to a comfortable level and well lit, and our water is heated, so that we can enjoy hot baths and showers at any time. The luxury of always being able to set the desired temperature is taken for granted today, one of the most pleasant achievements of our prosperous society. It is difficult to imagine the period after the Second World War, when not enough fuel was available in the cold winters to maintain even reasonably bearable temperatures in our dwellings. However, our prosperity and fossil fuels have eliminated these conditions once and for all – not everywhere in the world but at least for us (Figure 6.1).

But even eliminating fossil fuels will in no way jeopardize our privilege of always being able to select the temperature we want. Solar thermal energy – the heat from the sun – is an important alternative. Solar thermal energy covers a wide range of applications. When sunlight shines through windows, the power of the sun is already helping to warm up the building. For centuries, the sun has been a major source of heat for our homes.

In 1891 the metal manufacturer Clarence M. Kemp from Baltimore was awarded the first patent in the world for a technical solar thermal system. This was a very simply constructed storage collector for heating water. In 1909 the Californian, William J. Bailey introduced an optimized system concept that separated the solar heat collector from the water storage cylinder. Solar heating systems were marketed successfully in certain regions until the Second World War, after which the market collapsed because of competition from fossil fuels.

It was not until the oil crises in the 1970s that solar thermal power was rediscovered. In the years that followed there were still many teething problems, and not all systems ran smoothly. Today a variety of solar thermal system variants are available, and these systems are much more sophisticated than in the past. In combination with optimized heat insulation and other renewable heating systems, such as biomass heating and renewably operated heat pumps, these systems can make an important contribution towards the supply of carbon-free heat.

Renewable Energy and Climate Change, Second Edition. Volker Quaschning.
© 2020 John Wiley & Sons Ltd. Published 2020 by John Wiley & Sons Ltd.

Figure 6.1 Modern solar thermal collector systems are an important alternative to conventional oil and natural gas heating. Photos: http://www.wagner-solar.com.

> ### ⓘ Solar Collector, Solar Absorber, Solar Cell or Solar Module
>
> These terms are often used interchangeably, so to avoid confusion it is important to be clear about what exactly they refer to.
>
> A sun collector or solar collector is used to extract heat from the radiation of the sun. Thus, a collector always makes something hot. At the heart of a solar collector is a solar absorber. The solar absorber absorbs the radiation of the sun and converts it into heat. Solar collectors are used to heat domestic water, supplement space heating and produce high-temperature process heat. Thermal power plants can even generate electricity from high-temperature heat (*see Chapter* 7). However, even then solar collectors are first used to produce the heat.
>
> A *solar cell* is a photovoltaic cell that converts solar radiation directly into electric energy (*see Chapter* 5). Although a solar cell can also get hot, it is different from a solar collector in the sense that this is an undesired side effect. The heat actually reduces efficiency during the production of electricity. A solar module consists of a large number of solar cells and also generates electric energy from sunlight.

6.1 Structure and Functionality

Solar thermal energy is a technology for converting solar radiation into heat. The field of application for this technology is extensive. The higher the temperatures are required to be, the more complicated the technical implementation. The principle is similar with all solar thermal systems. A solar collector first collects the sunlight. The term collector comes from the Latin word 'collegere', which means to collect. The main component of a collector is the solar absorber. It absorbs the sunlight and converts it into heat (Figure 6.2). It then transfers this heat to a heat transfer medium.

The heat transfer medium may simply be something like water, air, or even oil or salt. Heat loss is unavoidable during the conversion. Part of the solar radiation is reflected

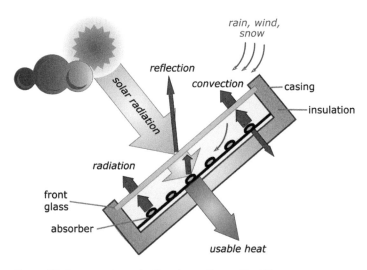

Figure 6.2 Processes in a solar flat-plate collector [Qua13].

and does not even reach the absorber. A certain amount of heat is lost before it can be transferred to the heat transfer medium. The trick is to construct a collector, so that heat loss is as low as possible, but the collector is still cost-effective to produce. Appropriate collectors should be used depending on the field of application and the desired temperatures.

The efficiency curve describes the performance of a collector. Research institutes measure collectors under strictly defined conditions to determine the curve. These characteristic curves can then be obtained from the manufacturer of the collector or over the Internet.

- www.spf.ch   Swiss collector test reports
- http://www.itw.uni-stuttgart.de   German collector test reports (in German)

Collector Efficiency

The efficiency and efficiency curve of a solar collector can be determined using the three parameters η_0, $a1$, and $a2$. The optical efficiency η_0 describes which portion of sunlight is converted into heat by the absorber. The absorber itself or the front panel of glass actually reflects a portion of the sunlight before it can even be absorbed. Depending on the type of collector, optical efficiency is between 70% and 90%. The two loss coefficients $a1$ and $a2$ indicate how high the heat loss is in the collector. The hotter the collector is, the higher the heat loss and the less useful heat the collector can emit. High loss coefficients also mean high heat loss. The formula for collector efficiency is as follows:

$$\eta = \eta_0 - \frac{a_1 \cdot \Delta\vartheta + a_2 \cdot \Delta\vartheta^2}{E}.$$

Here $\Delta\vartheta$ indicates the temperature difference between the collector and the environment and E the intensity of the solar radiation.

With an ambient temperature of 25 °C and a collector temperature of 55 °C, the result is a temperature difference of $\Delta\vartheta = 30$ °C or 30 K. On a pleasant summer's day with a solar radiation intensity of $E = 800$ W m^{-2}, the calculation of a flat-plate collector with an optical efficiency of $\eta_0 = 0.8$ and the loss coefficients $k_1 = 3.97$ W (m^2 K)$^{-1}$ and $k_2 = 0.01$ W (m^2 K^2)$^{-1}$ gives a collector efficiency of

$$\eta = 0.8 - \frac{3.97\frac{W}{m^2K} \cdot 30\ K + 0.01\frac{W}{m^2K^2} \cdot (30\ K)^2}{800\frac{W}{m^2}} = 0.64 = 64\%.$$

A collector 4.88 m^2 in size then supplies 2500 W of power. This is sufficient to heat 100 l of water from 33.5 to 55 °C in one hour.

The collector efficiency curve describes the relationship between the efficiency and the temperature difference between the collector and the environment (Figure 6.3). It shows that efficiency drops as the temperature rises until the collector reaches an efficiency of zero and zero output.

Almost all thermal solar systems require storage in addition to a collector. There are very few cases when the sun shines at exactly the same time as heat is needed. Even a simple water tank can function as a thermal store. It should be well insulated to reduce heat loss. The storage size depends mainly on the heat requirement and on how long the heat needs to be stored. Daytime storage cylinders for hot water systems in single-family homes are designed to bridge a few days of demand and usually only hold a few hundred litres. If, in addition to hot water, very large quantities of heat also need to be stored for heating, then seasonal heat storage cylinders are required. These store heat in the summer and then release it again in the winter. Large heat storage cylinders for small housing estates reach sizes of several hundred or even thousand cubic metres.

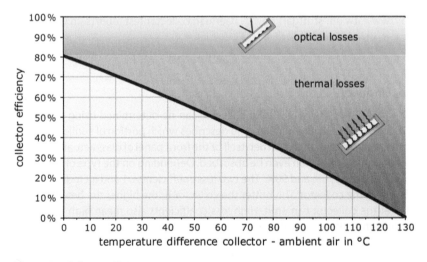

Figure 6.3 Collector efficiency curve.

Large storage cylinders generally have less specific heat loss than small ones. The storage volume increases considerably faster than the size of the storage surface. However, storage losses depend only on the size of the surface of the storage.

6.2   Solar Collectors – Collecting the Sun

6.2.1   Swimming Pool Absorbers

Different types of collectors are available for different purposes. The simplest type of collector consists of only one absorber. Placing a dark garden hose filled with water in the sun for a time produces enough heated water for a short but warm shower. A garden hose thus already has the characteristics of an absorber. In principle, a professional swimming pool absorber is also nothing more than a simple black plastic pipe that absorbs the sunlight almost optimally due to its dark colour. Weather-resistant plastics that cope well with UV light and chlorinated pool water are used for this purpose. The materials that are suitable are polyethylene (PE), polypropylene (PP) and ethylene-propylene-dien-monomere (EPDM). PVC should be avoided for ecological reasons. In winter it is futile trying to use a garden hose to extract warm shower water. The same applies to swimming pool absorbers. Although an absorber continues to absorb the sun's radiation, the heat loss in the absorber pipe itself is so great that hardly any heat can be drawn from the end of a pipe. Technically more sophisticated collectors are required to use the sun's heat during winter and the transitional periods of the year or when water temperatures should be warmer than the water in a swimming pool.

6.2.2   Flat-Plate Collectors

With flat-plate collectors a front panel of glass reduces heat loss considerably. Unfortunately, the front glass also reflects some of the sunlight. Therefore, when collector temperatures are very low, the efficiency of a swimming pool absorber can be higher than that of a flat-plate collector. However, flat-plate collector efficiency increases considerably as temperatures rise.

In summer, there are times when solar thermal systems do not need any heat. For example, when a heat storage cylinder is completely full the system does not pump any more water through the collector. This is referred to as a collector standstill. The temperatures in collectors can easily rise above 100 °C. Very good flat-plate collectors reach standstill temperatures between 150 and 200 °C. This therefore rules out the use of plastic pipes for absorbers. The absorbers of flat-plate collectors usually consist of copper or aluminium pipes fixed to a thin metal plate (Figure 6.4). The absorber itself is located in a collector enclosure that is well insulated at the back to minimize rear heat loss.

Metallic materials by nature do not have a black surface that absorbs the sun's radiation well, so they need to be coated. The first option that is usually considered for the coating is black paint. Although temperature-resistant black paint is good for this purpose, there are other, far better materials for absorber coating. When a black surface gets warm, it sends out some of the heat energy in the form of thermal radiation. This can be observed with an electric hot plate that has been switched on. One can feel the heat radiating from the hot plate without even touching it. The same effect occurs with

Figure 6.4 Cross-section of a flat-plate collector. Image: Bosch Thermotechnik GmbH.

a black-painted absorber. The absorber only radiates part of its heat to the water flowing through. Another part is emitted back into the environment as undesirable thermal radiation.

Thermal radiation losses can be minimized through the use of selective coatings (Figure 6.5). These coatings absorb the sunlight just as well as a black-painted plate. However, they emit a much lower amount of thermal radiation. Materials for selective coatings can no longer simply be painted or sprayed onto a surface. More complicated coating methods are now necessary.

6.2.3 Air-Based Collectors

In most cases, solar collectors heat water. However, for space heating what we actually require is warm air. With conventional heating systems, radiators or heating pipes installed under the floor transfer the heat produced from hot water to the air. Instead of water, air can be heated directly by a solar collector.

As air absorbs heat considerably less effectively than water, much larger absorber cross-sections are required. Otherwise, in principle, air-based collectors differ very little from liquid-based flat-plate collectors. Figure 6.6 shows an air collector with a ribbed absorber. An integrated photovoltaic module can deliver the electricity to drive the fan

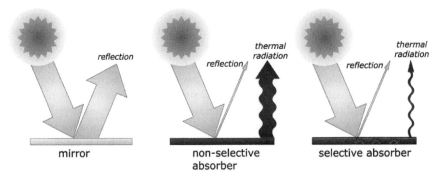

Figure 6.5 Principle of selective absorbers.

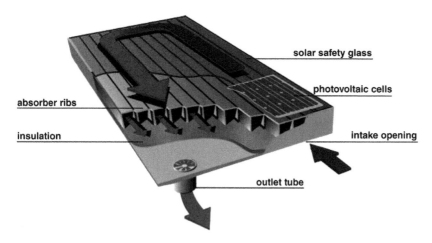

solar safety glass

photovoltaic cells

absorber ribs

insulation

intake opening

outlet tube

Figure 6.6   Cross-section of an air-based collector. Image: Grammer Solar GmbH.

motor. Air-based collectors are a particularly interesting alternative for auxiliary heating. However, heat storage is more difficult with systems that use air-based collectors.

6.2.4   Vacuum-Tube Collectors

With flat-plate collectors the air between the absorber and the front glass panel is the source of most of the heat loss. It causes convective heat loss and continuously transports the heat from the absorber to the glass panel. This then emits the heat unused back into the environment.

How an Inflated Plastic Bag Can Be Evacuated

The term vacuum originates from the Latin word 'vacus', which means 'empty' or 'free'. A vacuum is generally thought of as space empty of matter. However, it is practically impossible to produce a totally air-less space on earth.

In technology and in physics a vacuum is interpreted merely as air pressure that is considerably lower than normal air pressure. To produce a vacuum, one uses a vacuum pump to pump the air out of a stable space. In principle, it is also possible to use one's mouth to produce a rough vacuum – for example, by sucking the air out of a glass bottle. However, a plastic bag cannot be evacuated – at least not in any normal environment on earth. On the other hand, an almost perfect vacuum exists in space. If one opens an empty plastic bag in space and then closes it again, there will also be a vacuum in the blown-up bag. If one returns to earth with the bag, the ambient air pressure compresses it again.

Ambient air pressure originates through the power of the weight of air columns above the earth's surface. The atmospheric air weighs around $10\,t\,m^{-2}$ of earth surface. The question is, why doesn't this enormous pressure simply smash the glass of a collector? The answer is simple: the space under the glass is also filled with air, which produces the necessary counterpressure. On the other hand, if the air in the space behind the glass is evacuated, the glass will normally bend and shatter.

Figure 6.7 Vacuum tube collectors. Left: Collector with heat pipe. Right: Tube with direct flow. Images: Viessmann Werke.

Heat losses can be reduced considerably if a vacuum exists between the absorber and the front glass due to the air movement in the collector. This is the principle used by vacuum flat-plate collectors. As the outer air pressure would press the front covering against the absorber, spacers are needed between the underside of the collector and the glass covering. The vacuum cannot be kept stable for a long period because penetration of air where the glass and collector housing intersect cannot be completely prevented. Therefore, vacuum flat-plate connectors must be evacuated again at certain intervals. This is done by attaching a vacuum pump to a special valve on the collector. These drawbacks were the reason why vacuum flat-plate collectors never became popular.

Another type of collector available is the vacuum-tube collector, which does not have the same disadvantages as the vacuum flat-plate collector. With vacuum-tube collectors, a high level of vacuum is completely enclosed in a glass tube. Consequently, these collectors are considerably easier to produce and maintain for long periods than vacuum flat-plate collectors. The shape of the glass tubes enables them to better withstand outside air pressure so that metal rods are not necessary for support.

In the enclosed glass tube of a vacuum tube collector is a flag absorber, which is a flat metal absorber that has a heat pipe integrated in the middle (Figure 6.7 left). A lightly vaporizing medium like methanol is enclosed in this heat pipe. If it vaporizes due to the heat of the sun, the vapour rises upwards. The heat pipe sticks out of the glass tube at the top end. This is where the condenser, which condenses the heat medium again, is located. In the process, it transfers the heat energy over a heat exchanger to the water flowing through. After condensation the once-more liquid medium in the heat pipe flows downwards. For maximum functionality the tubes should be mounted with a certain minimum tilt.

With this system the heat carrier liquid flows directly through the collector. A heat exchanger is then not required, and the collector does not necessarily have to be mounted at an angle.

Because hydrogen molecules are extremely small, atmospheric hydrogen cannot be totally prevented from penetrating the vacuum. With time this also destroys the vacuum. So-called getters, which can chemically bind the hydrogen that penetrates the vacuum over a long period of time, are built into a collector to prevent this from happening.

The advantage of vacuum-tube collectors is a very high energy yield, especially during the cooler seasons of the year. Compared to one with normal flat-plate collectors,

Figure 6.8   Comparison of flat-plate and vacuum-tube collectors. Photo: Viessmann Werke.

a solar system with vacuum-tube collectors requires less space for the collector. The disadvantage is the substantially higher cost of the collector (Figure 6.8).

6.3   Solar Thermal Systems

6.3.1   Hot Water from the Sun

The sun can be used to heat water to a high temperature. This is something we learned from Archimedes, who used a convex mirror to bring water to the boil as far back as 214 BCE. That's not surprising, some will say, because Archimedes was not from Hamburg, London, or Vancouver. In Southern Italy it is simply easier to use the sun's energy. This is quite correct, but only up to a point.

Average temperatures in the Mediterranean region are much higher than in Germany, Great Britain, or Canada and there is no worry there about frost damage to hot water systems. Systems that use solar energy to produce hot water can be constructed more simply and thus more cheaply in frost-free regions than in colder climates. However, even at high latitudes solar energy is an excellent option for producing useful heat.

6.3.1.1   Gravity Systems

However, the fact that the use of solar energy should be virtually obligatory in southern countries has not yet spread everywhere. In other words, at similar latitudes the number of systems installed varies considerably between countries. Thermal solar systems are common in Cyprus, Turkey, and Israel but not in Spain or the southern USA, although these areas also have a sunny climate.

Gravity systems are mainly used in countries that do not get frost (Figures 6.9 and 6.10). A flat-plate or vacuum-tube collector collects the solar radiation and heats water that flows through the collector. A hot water storage cylinder is installed at a higher

Figure 6.9 Left: Demonstration model of a gravity system. Right: Gravity system in Spain.

level than the collector. As cold water is heavier than warm water, it sinks down from the storage cylinder to the collector. Here the water is heated by the sun and then rises to the top until it reaches the storage cylinder again. If no more solar radiation is available, the cycle stops until the sun starts it up again.

If properly designed, this kind of solar system can cover almost all the hot water needs of a family living in a warm, sunny, southern climate. It is only during some of the sun-starved weeks of the year that the water will not completely reach desired temperatures. Either this is accepted in those countries or a supplemental heater, for example, with

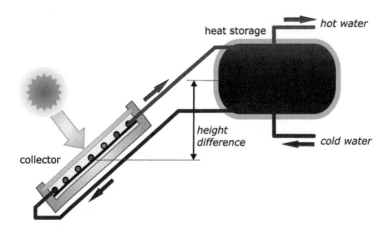

Figure 6.10 Solar gravity system (thermo-syphon system).

natural gas, is installed. If on occasion the heat of the sun is not sufficient, a supplemental heater then takes over.

In China solar thermal systems for heating water have gained a high market share. In remote rural regions there is almost no access to fossil fuels. The inhabitants of these regions do not have the possibility of heating their water as and when required if there is no sun. China uses mainly vacuum-tube collectors because of the need to guarantee high supply reliability from solar systems at low ambient temperatures.

6.3.1.2   Systems with Forced Circulation

A system should be technically optimized if the sun is to be used to produce hot water in areas with low outdoor temperatures. It would be too risky to heat up water for domestic use directly in a collector because the water would freeze in winter and therefore could destroy the collector. Consequently, the water that flows through the collector is mixed with an antifreezing agent. However, antifreezing agents cannot be used in the water supply, as they have negative effects on health. A heat exchanger therefore separates the water circulation from the solar circulation and transfers the heat to a hot water storage cylinder. The storage cylinder is normally designed so that it can provide enough hot water for two to three bad days until a major supply of heat comes from the solar collector (Figure 6.11).

Flat roofs are less common in Central and Northern Europe and North America than they are in Southern Europe. Hot water storage cylinders are traditionally located in a basement or in a utility room. If the hot water storage is situated below the collector, the water must be pumped through the collector. A pump moves the water in a solar circuit through the collector and a control mechanism ensures that the pump only starts

Figure 6.11   Single-family house with photovoltaic system (left) and flat-plate collectors for heating water (right). Photo: SunTechnics.

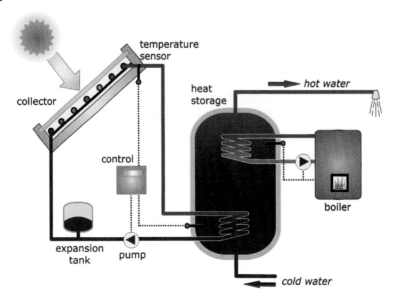

Figure 6.12 Pumped solar thermal system for heating domestic hot water.

when the temperature of the collector is higher than that of the storage cylinder (Figure 6.12). A conventional boiler heats up the water during the transitional seasons and in winter, so that a hot shower is possible all year round. It can happen that during the summer a solar collector will heat up an entire storage cylinder to a predetermined maximum temperature. This maximum temperature is usually set at 60 °C to prevent large deposits of lime from forming. If the cylinder is full, the control interrupts the incoming supply from the collector. Despite full irradiation from the sun, no more water will flow through the collector. The collector can then heat up to temperatures well over 100 °C and evaporate the water. If the expansion tank is large enough, it absorbs the expansion of the water volume.

6.3.2 Heating with the Sun

Solar thermal systems can be used to provide not only hot water but also heating. In principle, this only requires increasing the sizes of the collector and the storage cylinder and connecting them to the heating cycle. As no provision is normally made for domestic hot water in the heating cycle, two separate heat storage cylinders are needed – one for hot water and the other for drinking water. Combined-storage tanks can integrate both storage elements and reduce heat losses (Figure 6.13).

In places like Germany and Great Britain the sun is only able to meet heating demand during the transitional periods of spring and autumn. Consequently, these systems are usually designed so that solar energy only serves as a supplemental heat supply. The output of the collectors is not sufficient to cover the heat needed during the winter. In principle, the sun could be used to cover all heating requirements. However, this necessitates a very large storage cylinder that stores thermal heat from the summer for the winter and due to its size can usually only be integrated into new buildings. Although

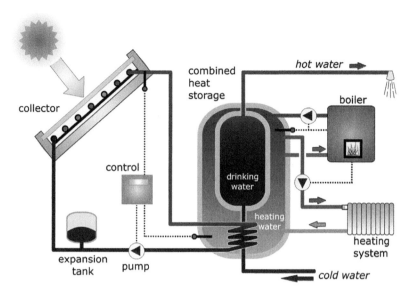

Figure 6.13 Solar thermal system for domestic hot water heating and auxiliary space heating.

this considerably increases the costs for the solar system, there is then no need for an additional heating system. For economic reasons, current practice tends to use tanks that can cover only a few days. Such systems enable between 20% and 70% of the heat demand to be covered by the sun (Figure 6.14).

Figure 6.14 Large roof-integrated solar thermal system for heating water and providing supplementary heating. Photo: SunTechnics.

Figure 6.15 In the energy self-sufficient solar house (left) in Lehrte, a 46 m^2 solar thermal system with a 9300 l buffer storage tank (right) covers 65% of the heat demand and an 8 kW PV system covers the entire electricity demand. Photos: HELMA Eigenheimbau AG.

Solar coverage rates of 20–30% are relatively easy to achieve for standard new buildings and even renovated old buildings. However, buildings with higher solar fractions are also needed in order to completely convert the heating sector to renewable energies for effective climate protection throughout Germany and elsewhere. To achieve this, the building must be optimally insulated and brought up to passive house standard. In a normal single-family house, 40–50 m^2 of collector surface and around 10 000 l of storage volume are then sufficient to achieve a solar fraction of 65%, even in less sunny locations. The remaining heating energy demand can then be covered quite easily with a wood-burning stove, for which 1–2 m^3 of firewood per year are sufficient. The annual heating bill is then usually less than €200 (Figure 6.15).

If the entire heating energy demand is to be covered by the sun, even larger collectors and storage tanks are required. Numerous successful solar houses in Germany and Switzerland have already proven that this is possible in principle. In a normal single-family home, the required collector size is around 80 m^2 and the storage tank size around 40 000 l.

6.3.3   Solar Communities

If many houses in a community have solar collectors, these can be integrated into a solar district heat network. This can also involve setting up a large central collector complex. At the heart of such a heat network is a central heat storage tank. Its size helps to minimize heat losses, thus also enabling heat to be stored for longer periods. However, the extensive tube systems involved can result in disadvantages, such as higher costs and the possibility of major line losses.

Several solar district heating systems have already been successfully installed. Europe's largest solar thermal project is currently located in Marstal, Denmark. A collector area of around 19 000 m^2 has been installed in the project, which has been implemented in various construction phases. A storage capacity of 15.5 million litres

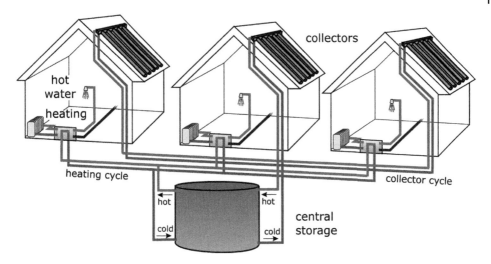

Figure 6.16   Solar community heating system.

is available through a combination of a storage tank, a storage pond and a gravel store. The solar heat is distributed via a local heating network and covers around 32% of the heat demand of over 1400 customers (Figure 6.16).

6.3.4   Cooling with the Sun

As paradoxical as it may sound, the heat from the sun can also be used to provide excellent cooling for buildings. In the hot and sunny regions of the world large numbers of energy-hungry air conditioning systems ensure that room temperatures are pleasantly cool. The sunnier and hotter it is, the greater the need for cooling. As the radiation from the sun increases, the output of a thermal collector also increases. In contrast with the requirement for heat, cooling load demand coincides almost perfectly with the supply of the sun.

Along with a large and efficient collector, an absorption-refrigerating machine is an essential element of a solar cooling system (Figure 6.17). In this context, the term absorption does not refer to a solar absorber. An absorption-refrigerating machine utilizes the chemical process of sorption. A chemist interprets sorption or absorption as the absorption of a gas or a fluid by another fluid. A popular example is the dissolution of carbon dioxide gas in mineral water.

Absorption-refrigerating machines use a cooling agent with a low boiling point, such as ammonia, which is later dissolved in water. Even aside from ammonia, water itself under low pressure can be used as a cooling agent. Lithium-bromide is then suitable as the solvent.

The cooling agent boils in a vaporizer at low temperatures. In the process it extracts the heat from the cooling system. The cooling agent must then be liquefied again so that it can provide continuous cooling through renewed evaporation. With the help of a few tricks and an indirect way through sorption, the liquidation also succeeds through the use of solar heat.

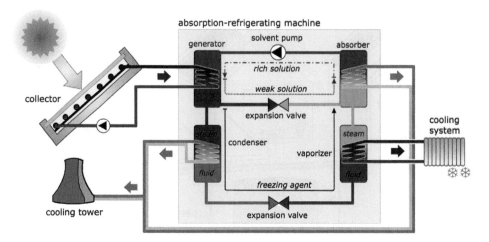

Figure 6.17 Principle of solar cooling with absorption chillers.

The cooling machine absorber first mixes the cooling agent vapour with the solvent. This produces sorption that releases the heat. This heat is either used to heat the water or is transferred to a cooling tower. The liquid solution, which has been enriched with a cooling agent, is transported to the generator by a solvent pump. The generator separates the cooling agent and the solvent on the basis of their different boiling points. Heat from efficient solar collectors is used for the boiling process. Temperatures of 100 –150 °C are optimal. The separated vaporous cooling agent then enters a condenser, which liquefies the cooling agent. The condensation heat is also dissipated either as useful heat or over a cooling tower. The liquid cooling agent enters the vaporizer over an expansion valve and the solvent returns again to the refrigerating machine absorber. The cooling agent cools down considerably through the expansion in the expansion valve, and can again deliver cooling to the cooling system over the vaporizer.

6.3.5   Swimming with the Sun

During the outdoor swimming season in Central Europe water temperatures normally reach 16–19 °C. The water is only warmer than this for a few days during the height of summer, although this period is becoming longer because of global warming. These water temperatures are a result of swimming pool water being heated by the sun. The following example shows how enormous is the energy content of the sun. The example uses Lake Constance, a lake popular for swimming that is visited by millions of tourists annually. In winter the lake cools off quite considerably and in 1880 and 1963 it was even completely frozen over. In summer, in contrast, it reaches a temperature high enough to attract hordes of swimmers. If the entire supply of coal consumed in Germany each year were used to heat the water, it would only be enough to heat up the 50 km^3 of water of Lake Constance just once to a temperature of 9 °C. Looking at the USA, the whole US American primary energy supply of one year would only suffice to heat up Lake Michigan by less than 5 °C on a single occasion.

The sun, on the other hand, can provide pleasant water temperatures without a problem and can do so over many weeks. Despite the heat of the sun, many bathers still find

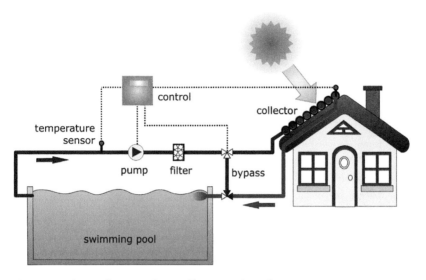

Figure 6.18 System for swimming pool heating using solar energy.

swimming pool temperatures too low in summer, so that pools are artificially heated. And because we only have to heat up a small swimming pool, and not all of Lake Constance or Lake Michigan, many places pull out all the stops. In Germany there are about 8000 public outdoor and indoor swimming pools. In addition, there are around 500 000 private pools. In the USA there are about 8.6 million pools in total and 270 000 commercial pools. Several hundred million euros are spent each year alone to heat public pools. Yet alternatives are available that would definitely enable us to use the sun to save fuel costs and considerably reduce the emission of carbon dioxide.

Outdoor swimming pools are particularly suitable targets for the use of solar energy. The popularity with bathers and the supply of solar energy are a perfect match. Simple swimming pool absorbers heat up pool water directly. A pump conveys the water through an absorber and a simple control ensures that water is only pumped when the sun actually can heat it up (Figure 6.18). If the water were pumped through the absorber tube at night as well, this would result in the pool water being cooled again. A good half-square metre of solar absorber surface is needed per square metre of pool area. The surface space that is needed is often found on buildings near the pool. Covering a pool at night can save further energy.

6.3.6   Cooking with the Sun

In many countries cooking is still often done on an open wood fire. About 2.5 billion people around the world use this traditional method to prepare their meals. From the energy point of view, however, an open fire is anything but efficient. Firewood does not last long. In many countries wood is cut down at a faster rate than trees can grow back. Furthermore, the smoke produced by open fires is responsible for many illnesses. In sunny countries solar cookers offer an alternative to traditional hearths.

A solar cooking box is a very simple cooking system: a wooden box painted black inside that is covered with a sheet of glass angled in the direction of the sun. This very

Figure 6.19 Solar cooker in Ethiopia. Photo: EG Solar e.V., http://www.eg-solar.de.

simple solar collector is actually capable of heating water and food. However, it is not very efficient, and the glass cover makes it difficult to prepare food. An efficient solar cooker is a neater solution. With a solar cooker, the cooking pot is situated in the middle of a convex mirror. The mirror is moved approximately every quarter-hour, so that it is directed optimally towards the sun. With a mirror diameter of 140 cm and good sun radiation, it is possible to bring 3 l of water to the boil in about half an hour (Figure 6.19).

6.4   Planning and Design

Of all the solar thermal systems described, the systems that use solar heat to heat up domestic water and to supplement other heating systems are the most widely used. The planning tips are therefore limited to these two variants.

6.4.1   Solar Thermal Heating of Domestic Hot Water

6.4.1.1   Outline Design

In Germany, Britain and other temperate regions, solar thermal domestic hot water systems are normally designed so that on a yearly average the sun covers 50–60% of the hot water demand. As the amount of sunshine in these regions fluctuates considerably during the year, a solar system can usually only cover the total hot water demand during the summer months. In the winter, the solar share can fall below 10% (Figure 6.20). A conventional heating system then has to cover the rest. In sunny regions like California and Southern Europe the solar share can easily reach more than 80%.

In order to further increase the solar fraction over the year, the expenditure would have to be increased significantly. Doubling the size of the system will not double the

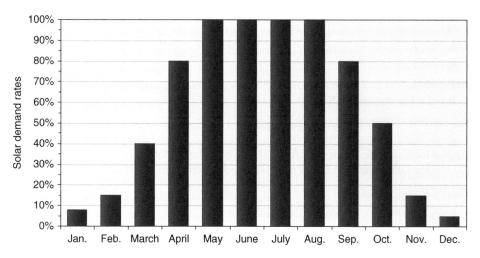

Figure 6.20   Typical solar fractions of solar thermal drinking water systems over the year.

solar fraction. Only in the summer months could the solar system meet double the demand. With the usual relatively small storage tanks, however, this surplus heat cannot be used. In winter, the share of solar energy would also increase. But if you double less than 10%, the share remains low.

A simple rule of thumb for designing solar thermal systems to provide domestic hot water, based on the number of people in a household, is:

- *collector size.* 1–1.5 m^2 flat-plate collectors per person; and
- *storage capacity.* 80–100 l per person.

If vacuum-tube collectors are used, the collector size can be around 30% smaller. The collectors should not be less than 3–4 m^2 in area, because below this size the losses in the tubes increase to above average.

6.4.1.2   Detailed Design

The hot water demand has to be determined before a detailed design can be drawn up. Ideally, the system should include a hot water meter that provides a direct reading of actual consumption or a record listing hot water use over a long period of time. As a rule of thumb, the demand can be approximated based on a hot water temperature of 45 °C:

- *Low consumption.* 15–30 l or 0.6–1.2 kWh per person per day
- *Average consumption.* 30–60 l or 1.2–2.4 kWh per person per day
- *High consumption.* 60–120 l or 2.4–4.8 kWh per person per day.

Storage Capacity and Hot Water Demand

The storage capacity, $V_{storage}$ should be about twice the total demand of P persons with a daily requirement V_{person} per person:

$$V_{storage} = 2 \cdot P \cdot V_{person}.$$

Based on the hot water demand Q_{person} per person per day, the annual hot water demand Q_{HW} for the supply of hot water can be calculated as:

$$Q_{HW} = 365 \cdot P \cdot Q_{person}.$$

The tank size for a four-person household with an average consumption of 45 l or 1.8 kWh of hot water per day would thus be:

$$V_{storage} = 2 \cdot 4 \cdot 45 \; \text{litres} = 360 \; \text{litres}.$$

Typical tank sizes are 300 or 400 l. A 400-l tank would be more than adequate in this case. A 300-l storage falls somewhat below the calculated demand but is still sufficient. The annual heating demand is:

$$Q_{HW} = 365 \cdot 4 \cdot 1.8 \; \text{kWh} = 2628 \; \text{kWh}.$$

■

Collector Size

Once the size of the storage cylinder has been established, the collector size $A_{collector}$ is calculated. This calculation also requires the annual figures for radiation H_{solar} and tilt gains f_{tilt} (cf. *Section 5.4*). With an annual solar fraction *sf* and an average system efficiency of 30% with systems using flat-plate collectors, the collector size $A_{collector}$ can be calculated as follows:

$$A_{collector} \approx \frac{60\%}{30\%} \cdot \frac{Q_{HW}}{H_{solar} \cdot f_{tilt}}$$

In this formula a solar fraction of 0.6 should be chosen for European climates (e.g. Berlin or London) and 0.9 for very sunny climates such as California. In this scenario, the collector size for flat-plate collectors should be calculated on the basis of a roof in Berlin orientated about 20° towards the south-southwest and tilted by 30°. The annual solar radiation in this case amounts to 1075 kWh $(\text{m}^2 \, \text{a})^{-1}$ and the southern orientation produces tilt gains of $f_{tilt} = 110\% = 1.1$. As a result, the required surface for flat-plate collectors is calculated as

$$A_{collector} \approx \frac{60\%}{30\%} \cdot \frac{2628 \; \text{kWh}}{1075 \; \frac{\text{kWh}}{\text{m}^2} \cdot 1.1} = 4.4 \; \text{m}^2.$$

■

The results, of course, depend heavily on the quality of the collectors and can vary considerably. Some online tools are available to help with system design (*see web tips*). Sophisticated computer programs are necessary to improve the detailed planning. Professional firms specializing in this field should also be able to provide detailed system designs. In addition to determining the size of the collector and the storage, their services include the design of other components such as pumps, controllers and pipes.

- valentin.de/calculation/thermal/start/de   Online calculations for solar
- www.solartoolbox.ch   thermal systems

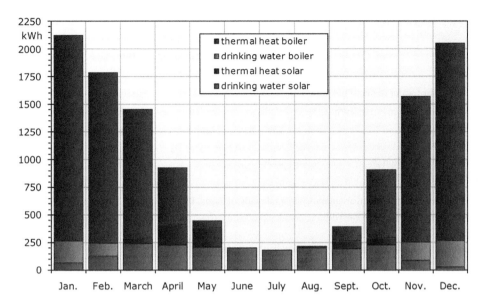

Figure 6.21 Typical monthly space heating and hot water demands in Germany and proportion of solar system versus conventional heating based on the demand of an old building with a total solar fraction of 20%.

6.4.2   Solar Thermal Auxiliary Heating

A large collector surface is needed if a solar thermal system is to provide support heating in addition to domestic hot water. In contrast to the supply of hot water, this option requires optimal building insulation to enable the sun to provide a larger share of the heat requirement. Whereas hot water use is relatively constant throughout the year, heating needs are concentrated in the winter months. However, the yield from solar collectors is low in the winter. Therefore, solar thermal systems that supply support heating are usually designed so that, in addition to hot water, they can cover only a portion of the heating required during the transitional period from March to October. In winter conventional heating systems essentially provide the heating required (Figure 6.21).

The size of the collector surface and the storage also affect the degree of solar coverage, and thus the proportion of heat covered by the sun. This also reduces the share a conventional heating system provides. If a fossil system fired with oil or gas is used, then the carbon dioxide emissions drop in accordance with the size of the solar system. However, a very large system also produces a higher surplus that cannot be used. Therefore, as a rule, large systems are less economical than small ones. So, when it comes to design, one must consider whether the priority is maximum input from the sun or economic viability.

The following two design options provide the basis for an outline design in Central European climates:

Option 1. Small system for good efficiency

- Collector area with flat-plate collectors: $0.8\,m^2$ per $10\,m^2$ living space;
- Collector area with vacuum tube collectors: $0.6\,m^2$ per $10\,m^2$ living space;
- *Tank size*: at least $50\,l\,m^{-2}$ of collector area.

Table 6.1 Solar fractions

	Existing building	New building pre-2010	Three-litre house	Passive house
Domestic hot water demand in kWh	2700	2700	2700	2700
Heating energy demand in kWh	25 000	11 500	3900	1950
Solar fraction, option 1 (small system)	13%	22%	40%	51%
Solar fraction, option 2 (medium-sized system)	22%	36%	57%	68%

Assumptions: Berlin location, orientation 30° south without shade, 130 m^2 living area, optimal flat-plate collector with combined-storage tank.

Option 2. Medium-sized system for higher solar fraction

- Collector area with flat-plate collectors: 1.6 m^2 per 10 m^2 living space;
- Collector area with vacuum tube collectors: 1.2 m^2 per 10 m^2 living space;
- *Tank size*: 100 l m^{-2} of collector area

An optimal design would of course also take into account the actual heating energy demand. The difference in requirements between an old building and an energy-saving three-litre house is considerable. Table 6.1 shows simulation results for optimal systems that were dimensioned according to the outline design described.

Although the collector with the medium-sized system is double the size of the small system and the storage is four times larger, this does not mean that the solar fraction is twice as high. The insulation standard of a building has considerably more influence on the solar fraction than these two elements. Optimal insulation options should be considered when the aim is to increase the solar fraction as a way of contributing towards climate protection.

 From Thinking about a Solar Thermal System on the Roof to Installing a System

- Determine the orientation and tilt angle of the roof.
    Is the roof suitable in terms of orientation and tilt angle?
    Recommendation: at least 95% according to Figure 5.15.
    Does the roof have too much shade?
- Decide whether only water for domestic use should be heated or whether solar heating or cooling support is also required.
- Outline design following rule of thumb.
- Possibly detailed design based on manual calculations or using simulation tools.
- Apply for appropriate approvals for listed buildings.
- Request quotations.
- Have the design specified by a specialist company.
- Apply for subsidies or low-interest loans (see following section).
- Award contract for the work.

6.5 Economics

6.5.1 When Does It Pay off?

Depending on type and design, between €200 and €350 m^{-2} should be estimated for flat-plate collectors and between €400 and €600 m^{-2} for vacuum-tube collectors in Central Europe. A 300-l heat storage cylinder costs around €700–1100. Installing a solar thermal system can be especially cost-effective in new buildings, or when an existing hot water storage cylinder is old and has to be replaced anyway. The costs for a solar thermal system fall within a wide range. A system purely for domestic hot water with 4 m^2 flat-plate collectors and a 300-l hot water storage cylinder can be acquired for as little as €2000, not including installation. The average cost of a European system for a four-person household excluding installation is between €3000 and €13400; a system including installation is around €5000. Depending on the size of the collector, the cost of a system that provides support heating can be double or even higher. Government grants are sometimes available to help cover the costs.

Even if the investment costs for a solar thermal system are known, getting a handle on the economics is difficult compared to photovoltaics (PV). The output of a photovoltaic system can be tracked accurately through an electricity meter and the statement for the fed-in electricity will show in euros and cents whether a system is living up to the planners' promises.

The output of a solar thermal system can also be monitored with a heat volume meter. However, in practice, these are hardly ever used because of cost. In very sunny countries a solar thermal system can cover the total demand. The cost-effectiveness is also usually high in these cases. In colder climates solar thermal systems are almost always supplemented by conventional heating systems, which make up for what the lack of sunshine cannot cover. No direct compensation is given for the heat output yield of a solar thermal system. A system only pays for itself indirectly through the savings in the fuel costs of the heating system. It is mainly the movement of fuel prices that establishes whether a system is paying for itself and at what rate. The higher fuel prices climb, the faster a solar thermal system will pay for itself. If, on the other hand, fuel prices fall, the economic viability of a solar system is less favourable.

Figure 6.22 shows the payback periods for a typical solar thermal system for heating domestic water. This is the time it takes for the fuel costs a system has saved to break even with the investment costs. If hot water is heated electrically for around €0.25 kWh^{-1}, the system will be pay for itself quite quickly. The payback becomes more challenging with particularly high-quality and expensive systems when fuel prices are low. In this case public grant programmes could be available to improve the economy of these systems.

6.5.2 Funding Programmes

In Germany, support for solar thermal systems is currently mainly provided through the Federal Government's market incentive programme, which is handled by the Federal Office of Economics and Export Control (BAFA). In the past, the funding rates were regularly adjusted to the number of applications. The application for funding should be submitted before construction begins. The process is relatively straightforward. The subsidy is paid directly, once the system has been installed.

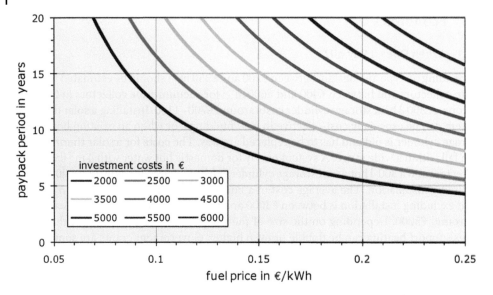

Figure 6.22 Payback periods for a solar thermal domestic hot water system with backup heating system (without interest rate effects and price increases; fuel savings: 2000 kWh a^{-1}, annual operating costs: 2% of capital costs).

In addition to support at federal level, there are also local support programmes in individual federal states. Low-interest loans, which are granted by the federally owned KfW development bank, can also be considered for some system options.

• www.solarfoerderung.de	Interactive support advisor
• www.bafa.de	German Federal Office for Economic Affairs and Export Control
• http://www.kfw.de	KfW development bank

6.6 Ecology

Solar thermal systems are among the most environmentally compatible renewable energy systems. They normally save on fossil fuels such as oil and natural gas, and thereby actively contribute towards climate protection. The collectors are mostly integrated into buildings and, consequently, do not require any extra land. The materials used in solar thermal systems – such as glass, copper, and plastic – are for the most part just ordinary materials commonly used in standard construction. Although environmentally problematic materials, such as polyurethane foam or PVC are also used in some solar collector systems, many collector manufacturers deliberately dispense with these materials.

Energy is needed to manufacture solar thermal systems. In Central Europe, it takes between six months and two years before a solar thermal system delivers the same amount of energy that was used in its production. This period is shorter in countries with a lot of sunshine. Many manufacturers place a great deal of emphasis on environmental protection and renewable energies during the production of thermal solar systems. The zero-emissions Solvis factory in Braunschweig, Germany, is an example.

Some attention needs to be paid to the size of electrical pumps in solar systems. These pumps require electric energy to operate. However, this requirement for auxiliary energy is usually smaller, by many orders of magnitude, than the energy saved by a solar system. Photovoltaic systems can be used to enable the auxiliary electric energy required by solar thermal systems to be covered directly by the sun.

Many solar thermal systems also use chemicals such as antifreeze or cooling agents. Typical antifreeze products like Tyfocor L have a minimal effect on water quality, and, therefore, are largely safe for the environment.

Special protective measures are required if ammonia is used for solar cooling in absorption-refrigerating machines. Ammonia is toxic and dangerous to the environment. Ammonia that escapes can bind to water. Lithium-bromide is also harmful to human health, but less so than ammonia.

6.7 Solar Thermal Markets

In terms of world markets for solar thermal collectors, one country puts all the rest to shame: China. Around 442 million square metres of collectors were installed in the country in 2015. These collectors deliver about 309 GW of heat power. The Chinese collector market thus accounts for over 70% of the entire world market (Figure 6.23).

With its widespread use of solar thermal energy, China has developed into a market leader in collector manufacturing. More than 1000 companies now produce and distribute solar collectors in China. Around 150 000 people were employed in the solar

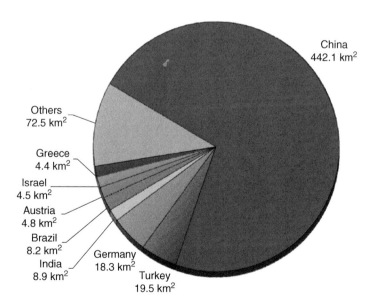

Figure 6.23 Installed glazed collector area in different countries. 2015. Source: Data: [SHC17].

thermal energy sector in the year 2000 in China [EST03]. In contrast to many other countries, China specializes in highly efficient vacuum tube collectors rather than the simple flat-plate collectors. Due to the large number of units produced for the Chinese market, Chinese suppliers are very competitive internationally. Most of the vacuum-tube collectors sold in the world come from China.

In the EU, Germany dominates the solar collector market. If one considers solar thermal collector area installed per head, small countries have the edge. In Cyprus, almost every house has a solar thermal system on it. In 2015 this Mediterranean island had around 679 000 m^2 of collector area distributed among just 1.1 million inhabitants. Statistically, this means there are nearly 606 m^2 of solar collectors per 1000 Cypriots. In Austria, by comparison, this figure is 595 m^2. Greece has around 400 m^2. Germany, on the other hand, only reaches 229 m^2 per thousand inhabitants and sunny France just 33 m^2. Local market conditions and acceptance by the population both have considerably more influence on the amount of installed collector area than the supply of solar radiation.

Germany is an example of how political conditions can affect markets. Until 2001 there was continuous market growth. In 2002 the market suffered a setback because of changes in incentive conditions and associated market insecurity (Figure 6.24). In 2008, the exploding oil price caused a brief boom in solar thermal energy, although this quickly evaporated again with falling oil prices. Since then, the solar thermal market has been under strong pressure due to low oil prices and is declining despite all the climate protection ambitions.

By the end of 2017, a total of 2.3 million solar thermal systems with a total area of 20.6 million square meters had been installed. In the same year the solar thermal sector in Germany accounted for over 10 000 jobs. The solar systems save the emission of around two million tons of carbon dioxide each year. Although this is an impressive number, it still falls far short of saving the climate.

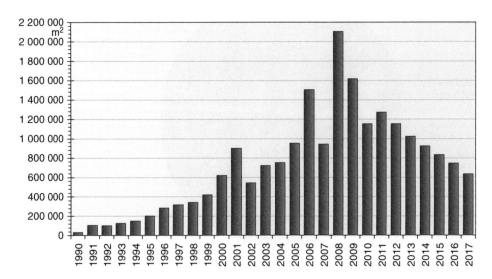

Figure 6.24 Annual newly installed collector area in Germany. Source: Data: [BSW18].

Figure 6.25 Many roofs still have space for solar collectors. Source: Photos: wagner-solar, http://www.wagner-solar.com

6.8   Outlook and Development Potential

During the past 15 years the solar thermal market in Europe has increased more than 10-fold. A few years ago, there were still quite optimistic estimates for future market development. According to these, the market for new installations was expected to increase to 8 million square metres per year by 2030 [BSW12] (Figure 6.25).

That would be around 10 times as much as in 2017. The number of employees in the solar thermal sector could then grow to over 50 000 in Germany alone. From today's perspective, however, this estimate seems too optimistic. While the costs for PV systems have fallen drastically in recent years, solar thermal energy has only been able to achieve comparatively small cost reductions and thus not achieve a major cost advantage over oil and gas heating.

In the meantime, solar thermal energy in Germany is in direct competition with photovoltaics. It could be that in the future a PV system will be installed on many houses

Figure 6.26 In some regions of the world, for example in southern Turkey, simple and thus inexpensive solar thermal systems already cover a large part of the heat demand.

instead of a solar thermal collector system and coupled to the heating system via a heating rod or a heat pump. Ultimately, it does not matter how solar energy is used for climate protection. The key is that our energy supply is completely covered by renewable energies as soon as possible. The potential is much greater in the sun-rich countries of the world. Due to the considerably higher economic efficiency, solar thermal energy could achieve a double-digit percentage share of the heat market there (Figure 6.26).

7

Solar Power Plants – Even More Power from the Sun

When we think of power plants, we usually imagine large central facilities with cooling towers and enormous chimneys emitting clouds of smoke. In terms of the concept itself, a power plant is merely a technical installation that converts a particular energy source into electricity. A solar power plant produces electric energy from solar radiation. We have already learned about one type of solar power plant – photovoltaic or PV systems. In principle, a small PV system sited on a single-family house does indeed constitute a solar power plant. However, since most people envisage a power plant as something much larger than a few square metres of solar modules, this chapter is devoted to large-scale systems that generate electricity from sunlight.

PV systems can in fact reach a size that is comparable with conventional power plants. Year after year new PV power plants have been breaking their own records in terms of size. PV systems with a capacity in the region of 5 MW and module areas of around 40 000 m^2 have already been installed on the roofs of industrial buildings. However, existing roofs are usually not appropriate for even larger systems. Plants with 50, 100, or even 1000 MW can only be achieved with ground-mounted systems. Large PV plants are technically very similar to the small systems on single-family houses, except that they are larger in scale. This chapter therefore deals with other types of large-scale solar power plants.

Concentrators can be used to increase the intensity of solar radiation. PV cells with very high efficiency then convert the concentrated sunlight into electricity, as this chapter will explain. In addition to PV, other technologies for generating power from sunlight also exist. Solar thermal power plants convert solar radiation first into heat and then into electric energy. A number of different interesting and very promising technologies are available and are particularly suitable for the very sunny regions on Earth because they need direct sunlight, which is far less abundant in temperate climates such as those of Northern Europe.

7.1 Focusing on the Sun

Most of us remember as children trying to set fire to a sheet of paper or a piece of wood using a magnifying glass. Even this simple experiment makes us aware of the energy that concentrated solar radiation can produce. On Earth, sunlight can, in theory, be concentrated by a factor of 46 211 and reach temperatures of 5500 °C at a focal point. In practice, concentration factors of over 10 000 and temperatures of well over 1000 °C

Renewable Energy and Climate Change, Second Edition. Volker Quaschning.
© 2020 John Wiley & Sons Ltd. Published 2020 by John Wiley & Sons Ltd.

Figure 7.1 Solar furnace near Almería in Spain. Large tracking mirrors direct the sunlight towards a convex mirror (top right) in the interior of a building.

have been reached. Large holes can easily be melted into steel plates using concentrated solar radiation in solar ovens. Solar ovens are also suitable for testing materials at high temperatures. Figure 7.1 shows a solar furnace near Almería in Spain, which reaches an output of 60 kW with full sun radiation.

For cost reasons, glass lens systems are normally ruled out for use as concentrators in large-scale technical applications. The reflector, which concentrates sunlight onto a focal point or at the focus, normally has the form of a parabola. Due to their long useful life, glass mirrors have proven to be reliable in practical applications. The reflector must be tracked, so that the sunlight always comes in at a vertical angle. A distinction is mainly made between single-axis and dual-axis tracking systems. Single-axis tracking systems concentrate the sunlight onto an absorber pipe at the focus point, whereas dual-axis tracking systems concentrate it onto a central absorber close to the focal point. The tracking can occur either over a sensor, which catches the optimal orientation to the sun, or through a computer that calculates the sun's position.

Archimedes – Inventor of the Concave Mirror?

The Greek scientist Archimedes of Syracuse, who was born in 285 BC, is considered the inventor of the concave mirror. Using concave mirrors, Archimedes is said to have set fire to enemy ships while his country was at war with Rome. The truth of this legend has been hotly debated for the last 300 years, but it is now considered unlikely. It would have taken very large and very precisely made mirrors, focusing on the ships over a very long period, to light the fires. Even if the legend is true, it did not do Archimedes much good. He was killed by the Romans when they captured Syracuse in 212 BCE.

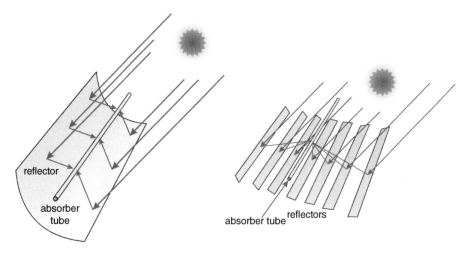

Figure 7.2   Single-axis tracking reflectors for line concentrators.

Parabolic trough collectors, which concentrate sunlight onto an absorber pipe, are normally used as line concentrators (Figure 7.2 left). A Fresnel collector is one in which the concentrator is distributed over multiple mirrors, each individually lined up in an optimal position towards the absorber pipe (Figure 7.2 right).

Concave mirrors are used for concentration onto a focal point (Figure 7.3 left). Distributed mirrors that are individually adjusted towards the sun can also concentrate the sun's radiation onto a central absorber (Figure 7.3 right).

7.2   Solar Power Plants

7.2.1   Parabolic Trough Power Plants

As the name indicates, in parabolic trough power plants large trough-shaped parabolic mirrors concentrate sunlight onto a focal point. The collectors are erected next to each

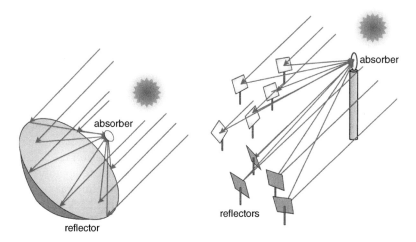

Figure 7.3   Dual-axis tracking reflectors for point concentrators.

Figure 7.4 View of the Kramer Junction parabolic trough power plant in California (USA). Photo: Gregory Kolb, SANDIA.

other in a row several hundred metres long (Figure 7.4). Many parallel rows in turn form an entire solar collector field.

The individual collectors rotate on their longitudinal axis and in this way follow the course of the sun. The mirrors concentrate the sunlight more than 80-fold at the focal point onto an absorber pipe. This is embedded in an evacuated glass casing in order to reduce heat loss. A special selective coating on the absorber pipe reduces the heat radiated from the pipe surface. With conventional systems, a special thermal oil flows through the pipe, which heats up to temperatures of about 400 °C as a result of the solar radiation. The heat is transmitted over heat exchangers to a water-steam cycle, vaporized and overheated again. The steam drives a turbine and a generator, which produces electric power. Behind the turbine it condenses again into water and then returns to the cycle through a pump (Figure 7.5). The principle of producing electricity using steam turbines is called the Clausius-Rankine process, named after its inventors. This process is also used in classic steam power plants, such as coal-fired plants.

During periods of bad weather or at night a parallel burner can also be used to operate the water-steam cycle. In contrast to PV, this guarantees a daily output of power. It also increases the attractiveness of and planning security in the public electricity supply. For totally carbon-free plant operation, either biomass or renewably produced hydrogen can be used as a supplementary fuel, or the burner can be eliminated entirely. Instead a thermal storage tank can be integrated. The solar field heats up the storage during the day using excessive heat. At night and during periods of bad weather the storage feeds the water-steam cycle (Figure 7.6). The storage must be designed to handle temperatures above 300 °C. Molten salt is suitable as a storage medium at this temperature range.

The development of solar thermal parabolic trough power plants dates back to 1906. In the USA and near the Egyptian city of Cairo – at the time still under British rule – research facilities were set up and the first tests were successful. Based on appearance,

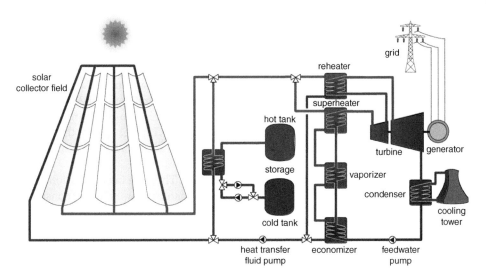

Figure 7.5 Parabolic trough power plant with thermal storage.

these trough facilities were incredibly similar to those used today. However, problems with materials and other technical difficulties ended this first attempt at large-scale technical generation of solar electric power in 1914, shortly before the start of the First World War [Men98].

In 1978 the foundations were laid in the USA for a resurrection of this technology. The Public Utility Regulatory Policies Act obligated American public electricity companies to buy energy from independent producers at clearly defined costs. After the surge in energy costs in the wake of the oil crisis, the electricity provider Southern California Edison (SCE) offered long-term feed-in conditions. The introduction of favourable tax advantages finally made the construction of plants worthwhile financially.

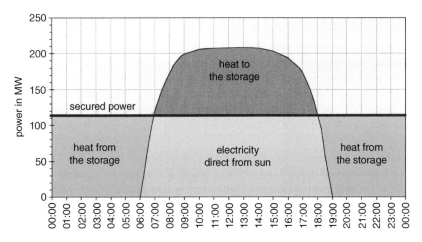

Figure 7.6 Solar thermal power plants with thermal storage can provide guaranteed output around the clock.

The LUZ Company, which was founded in 1979, negotiated a 30-year contract for the feed-in of solar energy with SCE in 1983. In 1984 the first solar thermal parabolic trough power plant was built in the Mojave Desert in California. By 1991, a total of nine SEGS power plants (Solar Electric Generation Systems) with 354 MW of electric output were installed in an area of more than 7 km^2. These power plants feed around 800 million kilowatt hours into the grid each year, enough to satisfy the requirements of 60 000 Americans. Eight power plants can also be operated with fossil fuels, thus enabling them to supply electricity at night or during periods of bad weather. With these plants the annual natural gas portion of the thermal energy supplied is legally set at 25%. The total investment for the plants was more than US$1.2 billion.

In the mid-1980s energy prices fell dramatically again. Then the tax exemptions also ran out at the end of 1990, and the LUZ Company went bankrupt even before construction could start on their tenth power plant. A long lean patch then followed for the planners of solar thermal power plants. It lasted until 2006, when building work was started on new parabolic trough plants in Nevada in the USA and near Guadix in Spain. Today, there are a number of new trough power plants in various countries.

Technical advances made in the meantime have helped to increase efficiency and reduce costs. One option was the use of direct solar steam. With this process water is vaporized at a high pressure by the collectors and heated up to 500 °C. This steam is led directly into a turbine, thereby rendering thermo oil and heat exchangers superfluous.

ⓘ Temperature and Efficiency

The Achilles heel of solar thermal power plants is not the solar collector that concentrates the sunlight. Collectors easily achieve efficiency of over 70%. However, most of the valuable solar heat is lost during its conversion into electricity. The steam turbines used in this process in solar power plants barely manage to achieve 35% efficiency. In other words, 65% of the heat gained from the sun returns unused as waste heat into the environment.

The efficiency of steam turbines results directly from the temperature difference of the steam between entry into and exit from the turbines. The exit temperature depends on the cooling and even at best is only minimally below the ambient temperature. With parabolic trough power plants, the entry temperature, contingent on the thermo oil, is currently just below 400 °C. With a temperature increase to 500 °C or higher, the efficiency of turbines could easily reach 40%. But this is the maximum even for steam turbines. Combined cycle gas turbine plants (CCGT), which operate at temperatures of over 1000 °C, achieve efficiency of up to 60%. CCGTs with high efficiency can be used, for example, in solar tower power plants.

One question that is often asked is whether simple pipe collectors can be used to generate electricity from general purpose water. In principle, this would be possible. However, due to the extremely low temperatures, efficiency would be too minimal to make this approach economically viable. Even the waste heat in the environment would have limited use. Solar power plants are normally situated in hot, sunny regions. However, a demand for gigantic quantities of low-temperature heat does not really exist in these areas.

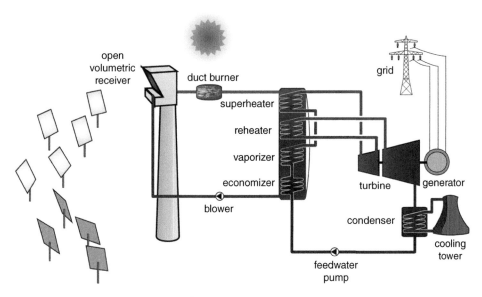

Figure 7.7 Solar tower power plant with open air receiver.

7.2.2 Solar Tower Power Plants

In solar tower power plants, hundreds or even thousands of rotating mirrors are arranged around a tower. Called heliostats, these mirrors are individually controlled by computer to track the movement of the sun and are orientated towards the top of the tower. They must be orientated precisely within a fraction of a degree, so that the reflected sunlight actually reaches the focal point. A receiver is located here with an absorber, which, due to the highly concentrated sunlight, heats up to temperatures of over 1000 °C. Air or molten salt transports the heat. A gas or steam turbine that drives a generator ultimately converts the heat into electric energy.

The tower uses open volumetric receivers (Figure 7.7), where ambient air is sucked by a fan through a receiver towards which the heliostats are orientated. The materials used for the receiver are wire mesh, ceramic foam, or metallic or ceramic honeycomb structures. The receiver is heated up by the solar radiation and transfers the heat to the sucked-through ambient air. The sucked-in air cools the front side of the receiver. Very high temperatures only develop on the inside of the receiver. This volumetric effect reduces radiant heat losses. The air, which is heated to temperatures between 650 and 850 °C, enters a waste heat boiler that vaporizes and overheats the water, thereby driving a steam turbine cycle. If required, this type of power plant can be fired by a channel burner using other fuels.

A more advanced version of the tower concept that uses a pressure receiver is now offering promising possibilities for the future (Figure 7.8). With this concept, the concentrated sunlight heats the air to temperatures up to 1100 °C in a volumetric pressure receiver at about 15 bars. A transparent quartz-glass dome separates the absorber from the environment. The hot air then drives a gas turbine. The waste heat of the turbine then finally drives the downstream steam turbine process. The first prototype has shown that this technology functions successfully.

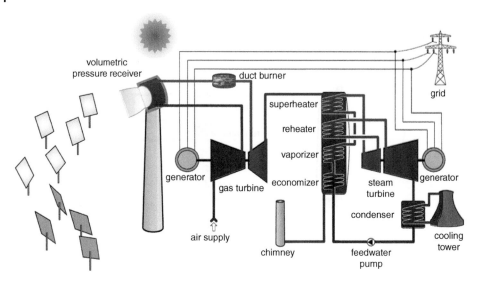

Figure 7.8 Solar tower power plant with pressurized air receiver.

With the combined gas and steam turbine process the efficiency of the conversion from heat to electric energy can be increased from about 35% to more than 50% with a pure steam turbine process. As a result, total efficiencies of more than 20% are possible with this type of conversion of solar radiation into electricity. These prospects justify the additional complexity and cost of receiver technology.

In contrast to parabolic trough power plants, there has not been much experience with commercial plants in the solar tower plant area. Research facilities that are optimizing system components and testing new components currently exist in Almería (Spain) (see Figure 7.9), Daggett (USA) and Rehovot (Israel).

The first commercial solar tower power plant to be put into operation was the 11-MW PS10 tower plant near Seville in Spain in 2006. However, instead of heating up

Figure 7.9 Research site for a solar tower power plant at Plataforma Solar de Almería (Spain).

air, the receiver of this plant vaporizes water. Due to the low temperatures it achieves, the efficiency of this power plant is still relatively low. In 2009, the 20 MW PS20 solar tower plant was completed near Sevilla. Several solar tower plants were planned and built in the USA. In contrast to the air-based concepts described above, these mostly use liquid salt as the heat transfer medium.

Before it can be launched successfully in the market, the open-air receiver technology developed in Germany first has to prove its suitability for practical use. This is currently being tested at a newly built solar tower plant in Jülich, Germany. With 1.5 MW of power this demonstration plant is considerably smaller than the commercial Spanish plants. The target in this effort is not the German power plant market but to promote the export of German technology to the sunny countries of the world.

7.2.3   Dish-Stirling Power Plants

Whereas trough and tower power plants are only economically viable in large-scale applications of many megawatts, the so-called Dish-Stirling systems can also be used in smaller units – for example, to supply remote villages or towns. With Dish-Stirling systems a convex mirror in the form of a large dish concentrates the light onto a focal point. To ensure that the light is concentrated as strongly as possible, the mirror is dual-axis tracked very precisely towards the sun.

A receiver is sited at the focal point. This receiver transfers the heat to the actual heart of the system: the Stirling hot gas engine. This engine converts the heat into kinetic energy and drives a generator that ultimately produces electric energy.

A Stirling engine can be driven not only by the heat of the sun but also through combustion heat. In combination with a biogas burner these plants can also generate electricity at night or during periods of bad weather. And the use of biogas also makes them carbon-neutral.

Some prototypes of pure solar systems have been built in Saudi Arabia, the USA, and Spain (Figure 7.10). Compared to tower and trough power plants, the price per kilowatt hour with Dish-Stirling systems is still relatively high.

Figure 7.10   Prototype of a 10-kW Dish-Stirling system near Almería in Spain.

7.2.4   Solar Chimney Power Plants

There is a big difference between solar chimney power plants and the thermal plants described earlier. While solar thermal power plants work by concentrating sunlight, solar chimney power plants function through the heating of air. The collector field is formed by a large flat area that is covered by a glass or plastic roof. A high chimney is located in the middle of the area, and the collector roof rises gently in the direction of the chimney. The air is able to flow in unencumbered at the sides of the enormous roof. The sun warms the air under the glass roof. This air then rises upward, follows the gentle slope of the roof and then flows at great speed through the chimney. The airflow in the chimney then drives wind turbines that generate electric power over a generator.

The ground under the glass roof can store heat, so that the power plant is still able to deliver power even after the sun has set. If hoses filled with water are laid in the ground, enough heat can be stored to enable the plant to provide electric power around the clock.

At the beginning of the 1980s a small demonstration plant with a rated power output of 50 kW was built near Manzanares in Spain. The collector roof of this plant had an average diameter of 122 m and an average height of 1.85 m. The chimney was 195 m high and had a diameter of 5 m. This plant was dismantled in 1988 after a storm knocked the chimney down. However, all the planned tests had been completed and the research plant lived up to expectations. It was the first successful demonstration of a solar chimney power plant.

Because the efficiency of solar chimney plants compared to other techniques is very low, these plants require large areas of land. Furthermore, the efficiency increases in line with the height of the tower, therefore, to be economically viable, plants must be of a certain minimum size. Similar power plant projects have been under discussion in Australia, for example. Consideration is being given to a large-scale 200 MW plant with a tower height of 1000 m, a tower diameter of 180 m and a collector diameter of 6000 m but so far, however, all projects have failed in the end due to funding issues (Figure 7.11).

Figure 7.11 Computer animation of a solar chimney power plant park. The towers can also be used as viewing platforms. Image: Schlaich Bergermann Solar, Stuttgart.

7.2.5   Concentrating Photovoltaic Power Plants

PV cells can also be operated by concentrating sunlight. The main point of this is that the concentration saves on valuable solar cell material. If sunlight is concentrated by a factor of 500, the size of the solar cell can then also be reduced by a factor of 500. The cost of the solar cell therefore becomes much less of an issue. This means materials that ordinarily would be too expensive without solar concentration could also be used. Concentrator cells, therefore, usually have higher efficiency than conventional PV modules.

There are many options for the concentration: for example, concentrator cells can be mounted at the focal point of parabolic troughs or convex mirrors. One of the main problems with this is efficient cooling, because, in addition to the electric energy of the solar cells, a large quantity of waste heat is produced. Flatcon technology takes a different approach. A flat Fresnel lens concentrates the sunlight onto a concentrator cell that is only a few square millimetres in size (Figure 7.12 below right). A copper plate on the back of the cell radiates the heat that accumulates extensively towards the back. A concentrator module comprises a large number of parallel cells. Many modules are then mounted together on a tracking device that orientates the modules optimally towards the sunlight.

7.2.6   Solar Chemistry

In addition to providing process heat or generating electricity, concentrating solar thermal energy can also be used for testing materials or in solar-chemical systems. For instance, there is a large solar furnace in the French town of Odeillo, where a large number of small mirrors are mounted on a hillside. These mirrors reflect the sunlight to a concave mirror 54 m in diameter, creating temperatures of 4000 °C in the solar furnace,

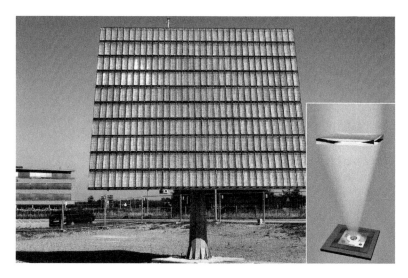

Figure 7.12 PV power plant with concentrator cells. Photo/graphics: Concentrix Solar GmbH.

which is used for research and industrial processes. Solar furnaces have also been built in Almería, Spain, and Cologne, Germany.

In addition to producing chemicals at high temperatures, solar thermal energy can also be used to produce hydrogen. This does not require a circuitous route that starts with generating electricity, followed by electrolysis. At high temperatures hydrogen can be produced through a solar-chemical process. For example, the chemical system could be in the receiver of a solar tower. Hydrogen is treated as an important energy source, particularly in the transportation sector and in fuel cells. If a hydrogen economy ever becomes a reality, concentrating solar-chemical systems could play a major role in hydrogen production.

7.3 Planning and Design

Solar thermal power plants are usually similar to typical large conventional thermal power plants. Due to their size, they cost millions of euros. Solar plants are almost always planned and built by large corporations or industrial companies. The design of these plants is usually very complex. It takes teams of engineers a long time to complete the detailed planning. One of the main goals is to optimize power plants from an economic perspective.

Private individuals are not likely to have any involvement in the planning of solar power plants, unlike the situation with small PV systems and solar thermal systems that heat tap water or provide auxiliary heating. However, the rise in the number of solar power plants is presenting many opportunities for investment in this area. Therefore, it is worth taking a quick look at the planning aspects.

Because concentrating solar power plants suffer a major reduction in efficiency in the partial load area, the aim should be to build them mainly in countries where there is a lot of sunshine. The regions currently of interest are those with an annual total global radiation of at least $1800\,kWh\,m^{-2}$. The optimal values are clearly those over $2000\,kWh\,m^{-2}$. These areas appear in red or pink in Figure 7.13.

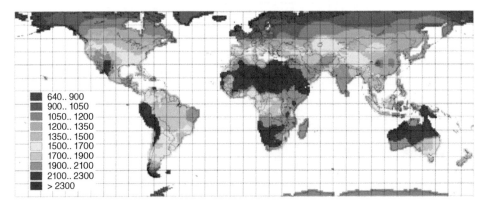

640.. 900
900.. 1050
1050.. 1200
1200.. 1350
1350.. 1500
1500.. 1700
1700.. 1900
1900.. 2100
2100.. 2300
> 2300

Figure 7.13 World map with annual totals for solar global radiation in kWh/m^2. Source: Meteotest, www.meteonorm.com.

7.3.1   Concentrating Solar Thermal Power Plants

Concentrating solar power plants can only utilize the direct irradiance portion of the sun. This portion of irradiance can be reflected by mirrors to be ultimately concentrated. The global irradiation intensity, i.e. the sum of direct and non-directional diffuse sunlight, is important for non-concentrating solar systems. The efficiency of these systems is therefore related to the global irradiation intensity. With concentrating systems, the direct-normal radiation intensity on a surface orientated vertically towards the sun serves as a reference quantity (see Figure 7.14). In the technical jargon this is abbreviated to DNI, which stands for 'direct normal irradiance'. The DNI values can lie somewhat below but also somewhat above the values for global radiation.

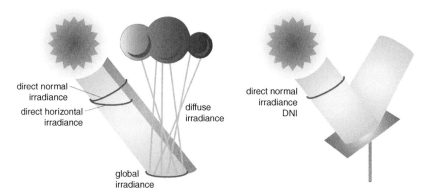

direct normal irradiance
direct horizontal irradiance
diffuse irradiance
global irradiance
direct normal irradiance DNI

Figure 7.14   Differentiation of types of solar radiation.

Annual Electricity Production of Solar Power Plants

The annual output yield of a concentrating solar thermal system can be estimated using the average efficiency, the annual total of solar direct-normal irradiance *DNI*, and the aperture area *A* of the mirrors:

$$E_{\text{electrical}} = \eta \cdot A \cdot DNI.$$

With an efficiency | of $15\% = 0.15$, a solar thermal parabolic trough system with an overall aperture area of $500\,000\,\text{m}^2$ at a Spanish location and a *DNI* of $2200\,\text{kWh}\,(\text{m}^2\,\text{a})^{-1}$ achieves an annual yield of:

$$E = 0.15 \cdot 500\,000\,\text{m}^2 \cdot 2200\frac{\text{kWh}}{\text{m}^2\,\text{a}} = 165\,\text{million}\frac{\text{kWh}}{\text{a}}.$$

This is sufficient to cover the electricity needs of around 50 000 Spanish households. ▪

Another interesting aspect of planning is how the land is used. So, that the mirrors do not throw shadows on one another, they can only be installed on about one-third of the available land area. The land area for a concentrating solar power plant must therefore be at least three times bigger than the surface area of the mirrors.

The mirror area should be coordinated optimally with the rest of the plant. If it is too small, the plant will constantly be running in partial-load operation. As a result, both efficiency and yield will drop. If the size of the mirror area is too large, a higher amount of solar radiation will fall onto the collectors than can ultimately be converted into electricity. In this case some of the mirrors have to be turned away from the sun.

The efficiency depends on the technology used. Dish-Stirling power plants and solar tower plants with pressure receivers can reach efficiencies of 20% or more. Solar tower or trough plants with steam turbines are currently at about 15% efficiency. An increase in DNI irradiance values also increases efficiency as a power plant will then be running for shorter periods in partial-load operation. In Germany, efficiency in the order of 10% is all that can be expected from its solar thermal power plants – and this is based on moderate DNI irradiance values of about $1000\,\text{kWh m}^{-2}$ and by year. The annual yield from concentrating solar power plants in places like Spain would be roughly three times higher than in Germany or Great Britain.

7.3.2 Solar Chimney Power Plants

With solar chimney power plants, the annual yield is calculated on an analogous basis. Instead of the DNI, the global radiation is used because these plants can also use diffuse sunlight. The total efficiency is only about 1% – and then only if the tower is more than 1000 m high. Thus, the efficiency of a solar chimney power plant is directly correlated to the height of the tower. At half the optimal tower height the efficiency is also halved. The collector area would have to be doubled in size to achieve the same yield. In contrast to a concentrating power plant, a solar chimney plant can use the entire land area. No unused gaps are necessary to prevent shading.

7.3.3 Concentrating Photovoltaic Power Plants

The annual yield from concentrating PV power plants is also calculated analogously. Compared to non-concentrating PV plants, an efficiency of more than 20% is possible. As concentrating PV plants can only use the direct-normal irradiance portion, the yield in sunny countries rises disproportionally.

7.4 Economics

Until a few years ago, solar thermal power plants were able to produce electricity much more cost-effectively than PV systems. Due to the strong cost reductions in PV, this advantage no longer exists today. In Germany, an economically viable operation of solar thermal power plants is practically impossible (Figure 7.15).

The main advantage of solar thermal systems is the simple integration of thermal storage. Small storage units used for a few hours only minimally increase the cost of generating electricity at solar thermal plants. However, they ensure that the power output of the plants can be guaranteed and, as a result, increase the availability and the value of electric energy. In solar thermal power plants, storage can generally be integrated much more cheaply than in PV. As a result, solar thermal power plants in very sunny countries are certainly of economic interest if high reliability is to be achieved through thermal storage.

In Spain, the cost of generating electricity from parabolic trough and solar tower power plants in 2012 was around 20 cents kWh^{-1}; at the top locations in North Africa

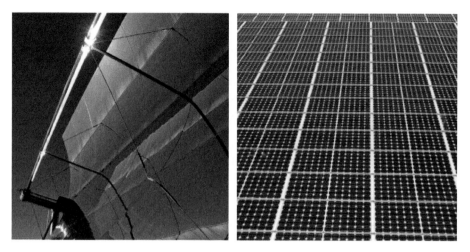

Figure 7.15 Solar thermal power plants (left) can achieve economic advantages over PV (right) when storage tanks are required for high reliability.

the cost was about 15 cents kWh^{-1}. As with PV, an increase in installed capacity of solar thermal power plants is expected to result in significant cost reductions. Costs have already been significantly reduced by the construction of a number of new solar trough and tower power plants. Meanwhile, the costs for new plants are in the range of 10 cents kWh^{-1}. In the future, costs in the order of 5 cents kWh^{-1} seem achievable.

In the case of solar chimney and Dish-Stirling power plants, there are no signs yet of a comparable rapid market development. Different studies show that these plants also have the potential to be competitive in electricity generation in the long term. However, it remains to be seen whether they can meet these expectations.

In the case of concentrating PV, electricity generation costs at locations with high solar radiation are also expected to be similar to those of conventional PV systems. Various suppliers have developed interesting systems in recent years. However, due to the strong cost pressure of standard PV, most suppliers of concentrating systems had to discontinue their activities.

The advantage of PV systems compared to solar thermal systems is modularity. PV systems can be built in any required size – ranging from milliwatts all the way to multi-megawatt plants several square kilometres in area. Solar thermal power plants are always dictated by the minimum output that is needed to make the plant economically viable. With Dish-Stirling power plants, this is around 10 kW. All other solar thermal power plants should have a minimum output of 10 MW. Efficiency is increased even further if output levels are between 50 and 200 MW. Optimally, a system should be 1 km^2 or more in total size to make it economically viable.

7.5   Ecology

Solar power plants without fossil fuel backup facilities do not release any direct carbon dioxide emissions during operation. When a fossil fuel-fired parallel burner exists, as is the case with some solar thermal parabolic trough power plants, the natural gas portion should be kept to an absolute minimum. Due to the low temperatures, the efficiency of solar thermal plants for electricity generation is lower than that of optimized pure natural gas plants. This aspect is insignificant when power plants are run purely on solar power. However, if fossil fuels are burnt as well, the carbon dioxide emissions rise. It is accepted that the addition of fossil fuels will increase supply reliability and protect against frost. But the fossil portion should not exceed 10% to ensure effective environmental protection.

Some World Bank projects in developing countries are striving towards an integration of relatively small parabolic trough collector fields into conventional gas and steam power plants run on natural gas. Due to technical restrictions, the solar portion is clearly below 10% there. This type of ISCCS (Integrated Solar Combined-Cycle System) is not suitable for effective climate protection.

The production energy required for thermal solar power plants is lower than that of conventional PV systems. Within a year a solar power plant will deliver more energy than was originally used to produce the plant.

Unlike small PV and solar thermal systems, solar systems cannot be integrated into buildings. Instead they need large open areas of space comprising many hectares. Ideally, solar plants should be erected in thinly populated areas where the land is not needed for other purposes. Fortunately, many sunny regions on Earth have just these characteristics. Very little grows in hot deserts where the sun shines almost all year round, and the human populations tend to be low. Solar thermal power plants built on suitable desert sites would easily be able to meet the electricity demands of the entire planet a hundred times over.

The main problem with thermal systems is the need for cooling water. Water is often a scarce resource in sunny regions. The small number of power plants installed until now have always been able to find local water reserves for cooling. On a larger scale, freshwater cooling in regions with little water is a problem. In principle, a solar thermal power plant can also operate with dry cooling. In this case, efficiency drops a little and the costs increase slightly. On a long-term basis, solar thermal power plants with dry cooling should also become economically interesting enough to resolve the water-cooling issue. If solar thermal plants are installed near the sea, the water from the sea can provide effective cooling. The waste heat of solar thermal facilities can also be used to desalinate sea water. It should be possible to produce carbon-free electric energy and drinking water at the same time, but until now this has not been done on a large scale.

7.6   Solar Power Plant Markets

The biggest markets for solar power plants exist in countries that have good solar radiation conditions and offer favourable energy credits. The largest installed solar power plant capacity is currently in Spain and the USA.

This is mainly due to the local economic conditions. Energy credits for solar thermal power plants in Spain are high and above the typical market prices for electric energy. The USA uses Renewable Portfolio Standards (RPSs). These vary from state to state and establish quotas for renewable energy plants. Plants are being built or planned in

Figure 7.16 Construction of a parabolic trough collector prototype in Andalusia. Spain is currently one of the largest markets for solar power plants alongside the USA.

certain US states such as California and Nevada. Other sunny and hot US states could follow suit.

• www.solarpaces.org	Information on solar thermal
• http://www.nrel.gov/csp/solarpaces/index.cfm	power plant projects

In the meantime, there are also larger solar power plants in Morocco, India, South Africa, Egypt, Mexico, and the United Arab Emirates. Further projects are underway in China, Australia, and Chile, among other countries. All large solar power plant projects are trough or tower power plants.

In Germany the focus has been mainly on developing the appropriate technology. Germany has been among the market leaders in the world, specifically in the area of components for solar thermal parabolic trough power plants. A prototype for a solar tower power plant, built in 2008 in Jülich, aims to showcase German technology and attract new export markets (Figure 7.16).

7.7   Outlook and Development Potential

Although the expansion in solar thermal power plants came to a standstill between 1991 and 2006, a few new facilities are now either at the planning stage or under construction. The current renaissance in the area of solar thermal power plants is likely to lead to further cost reductions. The main competitor of solar thermal power plants is PV, which is now much cheaper. At present, solar thermal power plants still have cost advantages in situations where storage is to be integrated. But there are also rapid cost reductions in PV systems with batteries. At the moment, therefore, it is not clear which technology will prevail in sunny countries when it comes to reliable electricity production with the aid of storage facilities.

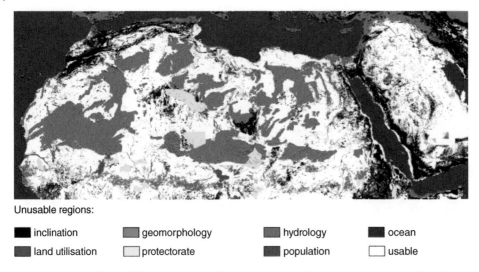

Unusable regions:

■ inclination		■ geomorphology		■ hydrology		■ ocean
■ land utilisation		☐ protectorate		■ population		☐ usable

Figure 7.17 Suitability of different regions in North Africa for building solar power plants. Suitable areas shown in white. Unsuitable areas colour-coded. Graphics: DLR.

There is enormous potential for the construction of solar power plants in North Africa. Even with a generous exclusion of unsuitable areas, such as sand dunes, nature reserves and mountainous and agricultural regions, about 1% of the remaining area of North Africa would theoretically be able to produce enough electricity to meet the demands of every country on Earth. Figure 7.17 shows the areas available there.

The question is, how can reasonably priced electricity from Egypt or Mauretania help us to solve our energy problems? The solution to this problem is simple. The cheap electricity would merely have to be transported to us. Figure 7.18 shows the top locations for obtaining this electricity and the possible transport routes to Europe.

Technically as well as financially, the transport of this energy is already viable today through high-voltage-direct-current transmission (HVDC). Transmission over a 5000 km long HVDC cable with losses of less than 15% is possible. These losses amount to around 0.5 cents kWh^{-1}, based on the possible electricity generation costs of 2–3 cents kWh^{-1} for PV and around 5 cents kWh^{-1} for solar thermal power plants. Added to this is the cost of the lines, which works out to between 0.5 and 1 cent kWh^{-1}. Altogether renewable energy could be produced at 3–6 cents kWh^{-1}, transported to Europe and, in combination with PV, wind or solar thermal power plants and integrated storage, guarantee high supply reliability.

This concept became known some time ago under the name DESERTEC and was promoted in 2009 with the participation of large corporations. In the meantime, many companies have since withdrawn. The Achilles' heel of the concept is the construction of long transmission lines, which is causing resistance in Germany and elsewhere, and the rather uncertain political situation in some North African countries.

Independently of the DESERTEC developments, some countries such as Morocco are in the process of establishing their own large solar power plant capacities. However, these primarily serve to cover the rapidly increasing electricity demand of their own country. If, in the future, solar power plant capacities reach dimensions where significant

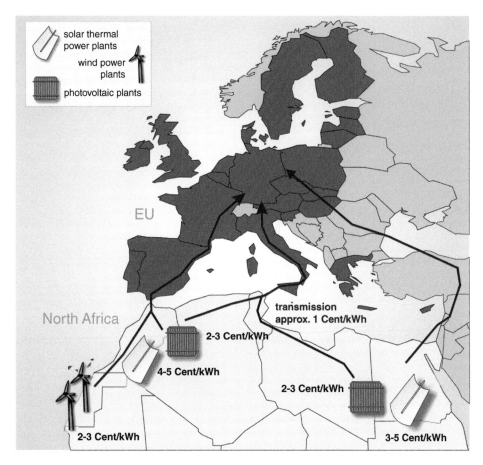

Figure 7.18 Options for renewable electricity imports from North Africa to the EU and electricity generation costs feasible in the medium term.

surpluses occur, they can certainly also be exported to southern European countries. The extent of new transmission lines required for this would be comparatively low. In the meantime, the costs for solar and wind power plants have fallen so sharply that they can also generate electricity in central Europe for well below 5 cents kWh^{-1}. It therefore remains to be seen to what extent an electricity import from Africa will contribute to the supply of electricity in Germany and neighbouring countries in the medium or long term.

8

Wind Power Systems – Electricity from Thin Air

With all the different renewable energies available today, wind power is considered a model technology because it has transformed from something of a cottage industry to a billion-dollar business within just a few years. During the mid-1980s wind power was still viewed as relatively unimportant, yet by 1995, 3655 wind turbines had been installed in Germany, with a total capacity of more than 1000 MW. By the end of 2017, Germany had 28 675 onshore wind turbines with an impressive capacity of 50 777 MW and 1169 offshore turbines with a capacity of 5387 MW, resulting in the creation of more than 100 000 new jobs.

Yet, the use of wind power is actually old hat and dates back many centuries. Years before the birth of Christ simple windmills were being used in the East for irrigation purposes.

Wind power was not exploited in Europe until much later. In the twelfth century post mills started to become popular for milling grain in Europe (Figure 8.1). These had to be rotated towards the wind, either manually or with the help of a donkey. In addition to milling grain, millers in those days also had the difficult job of managing the mill. When a storm approached, millers had to stop the mill in time so that they could remove the canvas from the vanes to prevent the power of the wind from destroying the mill. Wooden block brakes were used to stop the mechanism, but these were not without their own dangers. Sometimes the miller realized too late that he should have stopped the mill earlier. In their hurry to stop the vanes of the windmill from turning, millers would sometimes apply the brakes too hard and start a fire because of the heat that developed on the wooden brake blocks. This is why so many windmills burned down at the time. Long periods with no wind were another problem for millers, as a poem by Wilhelm Busch very aptly describes.

 Wilhelm Busch (1832–1908): Sod's Law (translated by Ros Mendy)

In his mill the miller's fretting.	Typical! the miller cries,
He's got forty sacks to fill.	this is driving me insane.
Outside though, the wind's abating –	When there's grain to grind, the wind dies,
the sails stop turning, all is still.	when it's windy, there's no grain.

In the centuries that followed, mills underwent considerable technical change, as the Dutch windmill shows. In addition to milling grain, they were now also used to pump

Renewable Energy and Climate Change, Second Edition. Volker Quaschning.
© 2020 John Wiley & Sons Ltd. Published 2020 by John Wiley & Sons Ltd.

Figure 8.1 Left: Historical post windmill in Stade, Germany. Photo: STADE Tourismus-GmbH. Right: Dutch windmill on the Danish island of Bornholm.

water and carry out other power-driven tasks. Technically sophisticated mills rotated automatically in the wind and could be braked safely. By the mid-nineteenth century 200 000 windmills were in use in Europe alone.

By the first half of the twentieth century steam-driven engines and diesel and electric motors had replaced almost all wind power systems. Wind power did not enjoy a resurgence until the oil crisis of the late 1970s. More and more sophisticated wind power plants are now offering a real alternative to fossil and nuclear power stations.

8.1 Gone with the Wind – Where the Wind Comes From

The sun is also ultimately responsible for the creation of wind. It bombards our planet with gigantic amounts of radiation energy, and the Earth needs to radiate this incoming solar energy back into space to save itself from continuous warming and ultimately burning up. However, more solar energy reaches the equator than the Earth radiates back into space. The path of sunlight to the South and North Poles is longer than to the equator. This explains why the situation at the poles is the exact opposite. Considerably less solar radiation reaches the poles and more energy is radiated into space than is emitted by the sun. As a result, a gigantic amount of energy is transported from the equator to the poles.

This transport of heat occurs primarily as a result of a global exchange of air masses. Major air circulations around the world pump the heat from the equator to the poles. Gigantic circulation cells, called Hadley cells, develop (Figure 8.2). The Earth's rotation deflects these currents. This in turn produces relatively stable winds that for a long time

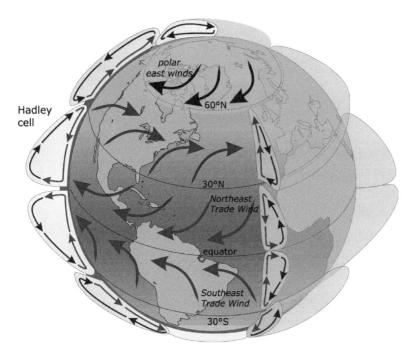

Figure 8.2 Global circulation and origins of different winds.

were crucial for sailing ships. In the tropical waters north of the equator a relatively stable and constant wind blows from the northeast. Unsurprisingly, it is called the Northeast Trade Wind. In contrast, the Southeast Trade Wind blows south of the equator.

Local influences exist in addition to the global wind currents. As a result of the rotation of the Earth, local areas of low and high pressure create wind movements that rotate around the low-pressure point. This rotation runs counterclockwise in the Northern hemisphere and clockwise in the Southern hemisphere.

Landward and seaward winds occur near the coast. During the daytime the radiation from the sun warms up the land, and the air above it, much more than the ocean water. The warm air above the land rises up, and colder air flows in from the ocean. At night this cycle reverses as land cools off faster than the ocean.

Headwinds also occur in the mountains and in polar regions with cold air streaming down mountain slopes sometimes at very high wind speeds.

Around 2% of solar radiation in the world is converted into wind movement. Therefore, the energy supplied by wind equates to a multiple of the primary energy demand of mankind. As is the case with water power, only a very small portion of this energy is usable. The largest supply of wind power is over the open seas, where no obstacles slow the wind down. Over land, the wind very quickly loses its speed due to the effects of rough terrain. For the same level of wind power as on the open seas it would be necessary to go to higher altitudes or use substantially larger tracts of land. Inland the effect on the wind due to the roughness of the terrain is no longer noticeable at altitudes higher than several hundred metres. As a result, wind use becomes more difficult as the distance

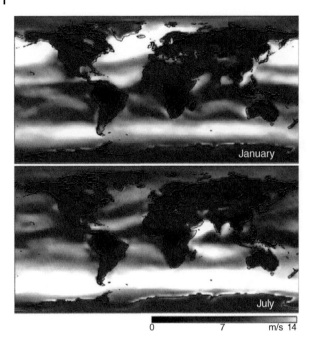

Figure 8.3 Average wind speeds worldwide. Source: NASA, http://visibleearth.nasa.gov.

from the coast increases. However, optimal locations also exist on hills and mountain tops inland.

The supply of wind energy on Earth varies considerably (Figure 8.3). Whereas it could just about cover the electrical energy demand in Germany, the potential in some countries is so vast that places like Great Britain could even export large quantities of wind power while covering all their own electric power needs. The best locations for wind power are usually where the wind hits land from the open seas without slowing down.

8.2 Utilizing Wind

Sailors have been using wind power as driving energy for centuries. However, one major characteristic of wind power is the strong fluctuation in the energy supplied. The capacity of the wind increases with the third power of wind velocity. Put more simply, this means that the capacity of the wind rises eightfold when wind speed doubles.

Power of the Wind

For a particular wind velocity v and density ρ, the power of the wind P_{wind} is determined by the area A:

$$P_{wind} = \frac{1}{2} \cdot \rho \cdot A \cdot v^3$$

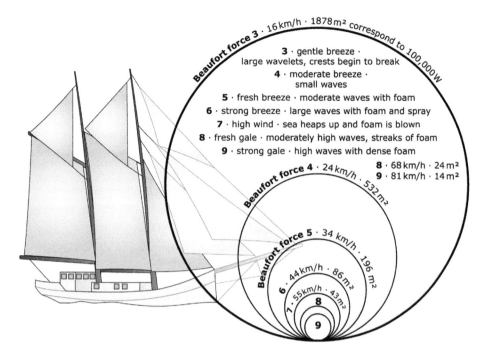

Figure 8.4 Area through which the wind reaches a power of 100 kW at different wind speeds.

With a wind velocity of 2.8 m s⁻¹, which equates to 10 km h⁻¹ or 6.2 mph, and an air density of $\rho = 1.225$ kg m⁻³, the power in the wind across an area of 1 m² is only

$$P_{wind} = \frac{1}{2} \cdot 1.225\frac{\text{kg}}{\text{m}^3} \cdot 1 \ \text{m}^2 \cdot \left(2.8\frac{\text{m}}{\text{s}}\right)^3 = 13.4 \ \text{W}.$$

With a wind velocity of 27.8 m s⁻¹, which equates to 100 km h⁻¹ or 62.1 mph, the power increases a 1000-fold and reaches 13.2 kW, which corresponds to around 18 hp (see Figure 8.4).

Meteorology still follows the practice of indicating wind velocity *v* based on the Beaufort scale (bft). The 12-level Beaufort scale was developed by the British Admiral Sir Francis Beaufort, who observed the sail behaviour of a naval frigate in different wind conditions and categorized it into different levels in 1806. The British navy officially introduced the Beaufort scale in 1838 (Table 8.1).

$$10 \ \text{m s}^{-1} = 36 \ \text{km h}^{-1} = 22.4 \ \text{mph}.$$

Systems that use wind energy must cope with a strong fluctuation in wind supply. On the one hand, they have to use the energy supplied by the wind even when wind speeds are low; on the other hand, they should suffer no damage even when wind speeds are extreme. Therefore, most wind turbines move into storm mode when wind velocity is very high.

Table 8.1 The Beaufort wind scale

bft	v in m/s	Description	Effect
0	0–0.2	Calm	Smoke rises vertically
1	0.3–1.5	Light air	Wind direction only detectable by smoke
2	1.6–3.3	Light breeze	Wind noticeable, leaves rustle
3	3.4–5.4	Gentle breeze	Leaves and small twigs in constant motion
4	5.5–7.9	Moderate breeze	Wind moves branches and thin twigs, raises dust
5	8.0–10.7	Fresh wind	Small trees begin to sway
6	10.8–13.8	Strong wind	Large branches in motion, overhead wires whistle
7	13.9–17.1	Moderate gale	Whole trees in motion, noticeable difficulty in walking
8	17.2–20.7	Fresh gale	Wind breaks branches off trees
9	20.8–24.4	Strong gale	Minor structural damage
10	24.5–28.4	Heavy storm	Wind uproots trees
11	28.5–32.6	Violent storm	Heavy storm damage
12	32.7	Hurricane	Devastation occurs

Wind Speed Records

The strongest gust of wind measured to date had a wind velocity of 412 km h^{-1} (114 m s^{-1}, 256 mph) on 12 April 1934 on Mount Washington in the USA. The highest 10-minute average of 372 km h^{-1} (231 mph) was recorded on the same day on Mount Washington. Germany's highest wind speed of 335 km h^{-1} (93 m s^{-1}) was measured on the Zugspitze mountain on 12 June 1985. Tornados reach even higher wind speeds. Radar has established wind speeds of around 500 km h^{-1} (139 m s^{-1}, 311 mph) with the strongest tornados recorded thus far. At this rate the wind capacity per square metre totals more than 1.6 MW or over 2000 hp. This same effect would be achieved if four large lorries of 500 hp each were driven at full speed towards a one square metre surface. On this basis it is not difficult to understand the extremely destructive effect of tornados.

Modern wind turbines use part of the kinetic energy of the wind. In so doing, they slow down the wind. However, it is physically not possible to use all the power contained in the wind, because this would require the wind to be stopped completely. This was something the German physicist, Albert Betz recognized in 1919, when he described how maximum energy can be extracted from the wind when it is slowed down to one-third of its original speed. According to Betz's law, no turbine can capture more than 16/27 (59.3%) of the kinetic energy in the wind. This is known as Betz's coefficient or Betz limit.

This power coefficient therefore indicates how much wind energy a wind turbine can utilize. Under ideal operating conditions, optimized modern wind turbines can use just over 50% of the energy contained in the wind and convert it into electric energy. They therefore reach power coefficients of at least 50%, close to the maximum possible limit. The symbol is cP. The power coefficient is a typical parameter for wind turbines and is also found in the data sheets provided by manufacturers.

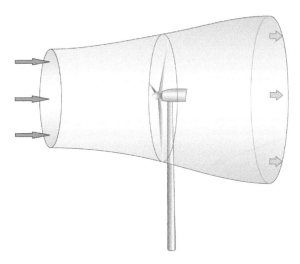

Figure 8.5   Flow profile of a wind turbine.

When a wind turbine slows the wind, it ends up changing the course of the wind flow. Behind the turbine the slowed wind flows through a large area and the wind flow widens (Figure 8.5).

When using the wind, a distinction is made between:

• Resistance principle; and
• Lift principle.

For example, the old Viking ships that were equipped with square rigs worked on the resistance principle. The sail put up resistance to the wind and the wind pushed the sail, propelling the ship forward.

With the lift principle, which is also used by modern sailboats with fore-and-aft sails, considerably more energy can be drawn from the wind than with the resistance principle. Consequently, modern wind turbines function almost exclusively on the basis of the lift principle. Among the different designs, the familiar rotors with a horizontal axis have been the most successful.

With these turbines, the wind flows from the front towards the rotor hub. Due to the relatively fast-rotating rotor blades, there is an airstream in addition to the actual wind and this airstream flows from the side onto the rotor blade (Figure 8.6). The rotor blade is then struck by a resulting wind that is composed of both the actual wind and the airstream.

The rotor blade now divides the wind and the resulting wind flows along the rotor blade. Due to the shape of the rotor blade, the wind has a longer distance to cover on the upper surface of the blade than on the lower surface. As a result, the flow widens and the air pressure decreases. Low pressure is created on the upper surface and high pressure is created on the lower surface. This difference in pressure then produces a lift force that reacts vertically to the resulting wind.

This lift force can be split into two components: the thrust force in the direction of the rotor axle and the tangential force in the direction of the circumference. Unfortunately,

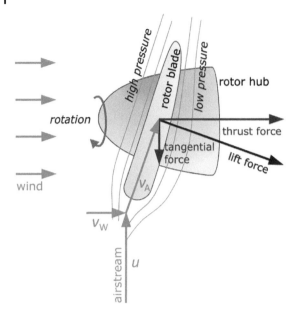

Figure 8.6 Functional principle of a wind turbine with horizontal axis.

in most cases the thrust force is bigger than the tangential force. The thrust force is not very useful. It merely presses against the rotor blades and bends them. The construction of the rotor blades themselves must therefore be very stable so that they can resist the thrust power. Ultimately, the tangential force is responsible for the rotation of the rotors because it works along the circumference.

Modern wind turbines always try to optimize the volume of tangential force. This requires an optimal relationship between the actual wind velocity and the airstream. The rotation of the rotor affects the airstream. The rotation of the rotor blades enables an optimization of the angle at which the resulting wind approaches the blade. This deliberate rotation of the rotor blade is known as 'pitching'.

8.3 Wind Turbines and Windfarms

8.3.1 Wind Chargers

One of the applications for smaller wind power systems is charging battery systems. Technically, an island system with a battery functions in a similar way to a photovoltaic island grid, except that a wind generator is used instead of a photovoltaic module. Small wind power systems are particularly popular for use on boats (Figure 8.7), where they are often used to charge on-board batteries for anchored craft.

Compared to photovoltaic island systems, island grids with wind turbines are technically more complex in their operation. If a battery is fully charged, the charge regulator of a photovoltaic system removes the photovoltaic module from the battery in order to prevent overcharging. Photovoltaic modules can cope with this without any problem. But it is a different case with wind generators that are only separated from the battery. If a wind generator does not have a load or a battery attached, high wind speeds can

Figure 8.7   Small wind turbines used to charge battery systems.

make the rotor blades revolve so fast that the generator is damaged. Therefore, some wind charge regulators switch wind generators with a fully charged battery to a heating resistance, thus restricting the number of revolutions.

The electric generators used by wind turbines are almost always alternating current or three-phase generators. However, a battery can only be charged with direct current. A rectifier converts the alternating current of the wind power system into direct current (see Figure 8.8).

Small wind turbines are usually mounted onto very low masts. The use of wind is usually not viable at sites that are far inland or where many trees, houses or other objects will obstruct its flow. Building regulations should also be checked beforehand to ensure that a wind system can be erected.

Storms generally pose a threat to wind generators. Large wind turbines therefore incorporate special storm settings. For cost reasons, small wind turbines often do not have any built-in storm protection. As a result, it is wise to dismantle the wind generator as a precautionary measure when large storms are forecast.

In addition to simple wind chargers, more complex island grid systems are available. These allow wind turbines to be combined with other renewable energy producers. In many cases, photovoltaic modules and wind generators work relatively well together. When there is too little sunshine, there is usually a strong wind or vice versa.

8.3.2   Large, Grid-Connected Wind Turbines

Large wind turbines that feed into the public electricity grid have seen extensive technical development during recent years. In the 1980s the typical capacity of a wind system was 100 kW or even less. The size of the rotors was still very modest at a diameter of less than 20 m (65 ft). In 2005 manufacturers were already developing prototypes for systems

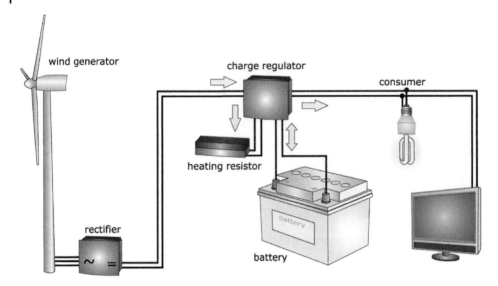

Figure 8.8 Principle of a simple stand-alone wind system.

with 5000 kW capacity, in other words 5 MW, and rotor diameters of 110 m (360 ft) and more. Today these systems are ready for mass production. Compared to the size of the systems available today, the traditional windmills and sailing vessels seem like toys (Figure 8.9). They even dwarf large airliners such as the Boeing 747, which has a wingspan of about 60 m (195 ft), and the mega-airliner A380 with its 80 m (261 ft) wingspan.

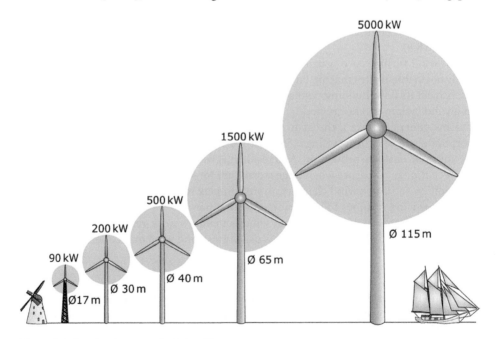

Figure 8.9 Size development of wind turbines.

Figure 8.10 Erection of a wind turbine. Top left: Foundation. Right: Tower. Below left: Rotor and nacelle. Photos: German Wind Energy Association and ABO Wind AG.

Apart from their size, the performance of modern wind turbines is also impressive. For instance, a single 6 MW system can supply all the electricity requirements for more than 5000 households in industrialized countries like Germany and Britain.

However, there are physical limitations to increasing the size of wind turbines. As the size increases, there is a disproportionate increase in the material requirements. Furthermore, there are major logistical problems in transporting building parts for such gigantic structures. Based on current conditions it is therefore highly unlikely that turbines of over 10 MW will be built.

Wind power is currently one of the most effective technologies for combating climate change. Even the largest wind turbines can be erected in just a few days. A concrete foundation provides a secure base (Figure 8.10). In places with soft subsoil posts are used to support the foundation in the ground.

There are three different types of towers that can be used: tubular steel towers, lattice towers and concrete towers. Years ago, lattice towers similar to electricity pylons were normally used. For aesthetic reasons tubular steel towers then became more popular. Because of rising steel prices and growing problems in transporting the large tubular segments needed for the biggest types of turbines, concrete, and lattice towers are now being used more and more. As the size of the turbines grew, so too did the demands placed on crane technology. Today cranes are required to lift many tons of heavy loads to heights of more than 100 m.

At the heart of modern wind turbines is the nacelle (Figure 8.11), which is mounted on the tower in such a way that it can rotate. A wind measurement device establishes

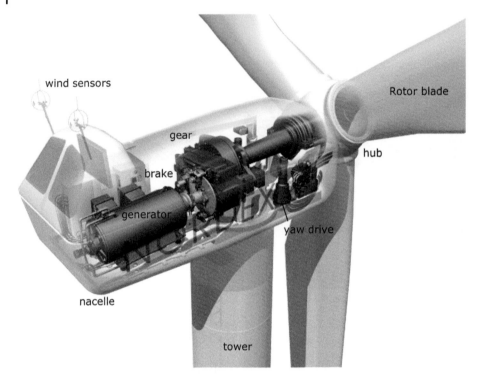

wind sensors

gear

brake

generator

nacelle

Rotor blade

hub

yaw drive

tower

Figure 8.11 Structure and components of a wind turbine. Graphics: Nordex AG.

wind speed and direction. The yaw drive then turns the turbine so that it faces the wind in an optimal position. The individual rotor blades are attached to the hub. A shaft starts up the movement of the rotor blades and powers the gears and the electric generator. The purpose of the gears is to adapt the slower rotation speed of the rotor to the faster rotation speed of the generator. Although the rotor of a 5 MW turbine with a diameter of 126 m only reaches a rotation speed of a maximum of 12 rpm, the rotor blade tip moves at speeds of over 280 km h^{-1}.

Some manufacturers are also promoting wind turbines that do not use gears. Whereas the electric generator of a wind turbine with gears operates at a rotation speed in the order of 1000 rpm, a gearless generator rotates at the rotor rotation speed. Depending on the size of the wind turbine, this equates to between 6 and 50 rpm. A generator must be larger in size to use the relatively slower rotor rotation rate. The savings on the gearbox are thus cancelled out because of a considerably heavier and more expensive generator. However, until now both wind turbine concepts, that is, with or without a gearbox, have been equally successful in the turbine market.

Modern wind turbines adapt their rotation speeds to the wind speed. When wind speeds are low a turbine reduces its rotation speed, thereby making better use of the wind. Wind speeds of 2.5–3.5 m s^{-1}, that is, 9–13 km h^{-1}, are usually sufficient to start a turbine. Wind turbines reach their maximum capacity with wind speeds of around 13 m s^{-1} (47 km h^{-1} or 29 mph). If the wind speed increases more than that, a turbine begins to limit its speed. It pitches the rotor blades less favourably into the wind, thereby reducing the momentum. This enables the turbine to keep its output at a constant level.

Figure 8.12 Maintenance work on wind turbines. Source: REpower Systems AG, Photos: Jan Oelker, caméléon und Stéphane Cosnard.

When wind speeds are very high at more than 25–30 m s^{-1} (90–108 km h^{-1}), a turbine moves into storm mode. The yaw drive turns the entire turbine out of the wind and brakes hold the rotor firm.

As large wind turbines already produce high output, they usually feed their electric current directly into a medium-high voltage system. A transformer converts the generator voltage into the mains voltage (see Figure 8.12).

Maintenance on wind turbines often requires an arduous climb for the technicians, so some turbines are equipped with lifts. Modern wind turbines have a relatively low incidence of outages, and a turbine service life of at least 20 years can be expected.

8.3.3 Small Wind Turbines

While large wind turbines in the megawatt range for feeding into the grid are largely mature, small wind turbines still only play a niche role. In principle, wind turbines can also be erected in cities. However, large turbines have too much impact in an urban setting, so that they are usually out of the question. If small wind turbines are mounted on low masts, they only achieve modest yields, especially in urban areas due to 'shading' from buildings and trees. The wind supply on roofs of tall buildings is slightly better. However, retrofitting wind turbines on roofs is not always easy. The structure of the building must be able to absorb the additional loads. Also, damage to the roof skin by the wind turbine must be avoided, in order to prevent moisture ingress. In many regions, the construction of small wind turbines directly on buildings or at ground level with mast heights of less than 10 m does not require a building permit. Irrespective of this, general guidelines for the construction and connection of wind turbines must be taken into account, and a structural analysis must be carried out for buildings.

Small wind turbines on masts usually have the classic design with horizontal axis. Other rotor profiles are often used in rooftop systems (Figure 8.13 right). Systems with vertical axis do not require alignment. They run more quietly and start at lower wind speeds. However, they often have somewhat poorer coefficients of performance and are

Figure 8.13 Small wind turbines at HTW Berlin.

therefore not able to exploit the wind as well as turbines with a horizontal axis. Compared to large wind turbines, the specific costs per kilowatt of installed capacity are considerably higher, and the yields are usually significantly lower. Economic operation of grid-connected small wind turbines has therefore barely been possible in Germany in recent years.

8.3.4 Windfarms

During the pioneering days wind turbines were often erected individually, whereas today they are almost always grouped in large windfarms (also known as wind parks). A windfarm consists of at least three turbines but can have many more. For example, the Alta windfarm erected in 2012 in California has 390 wind turbines with a total capacity of 1020 MW. These wind turbines are capable of supplying electricity to 200 000 US households. The Gansu windfarm on the edge of the Gobi Desert in China is expected to have an output of 20 000 MW by 2020.

The main advantage of windfarms compared to individual turbines is the cost saving. Planning, erection, and maintenance are considerably more economical. Large wind turbines usually have to have obstacle marking for air traffic. This includes a coloured mark on the rotor blade tips and navigation lights that come on when visibility is poor. In windfarms, only the outer turbines have a marking, which saves money and improves their appearance (see Figure 8.14).

The disadvantage of windfarms is mutual interference. If the wind turbines are sited close together, they can take wind away from each other. The efficiency of the turbines at the back is then impacted. It is important that sufficient distance is maintained between the turbines in the main direction of the wind so that efficiency losses can be minimized. However, efficiency loss through mutual interference cannot be totally prevented. Windfarm efficiency takes into account losses from mutual interference, which

Figure 8.14   Windfarms. Source: REpower Systems AG, Photos: Jan Oelker.

is usually between 85% and 97%. This means that losses between 3% and 15% should be expected.

If wind turbines are too close to housing estates, they can have a negative effect due to noise emission and a shadow being cast as the sun goes down. Turbines should ideally be erected several hundred metres from the nearest dwellings to avoid creating this effect.

Because of the rules on required distances, it is already becoming increasingly difficult to find good sites for new windfarms in Germany and Denmark. Although there is still considerable potential for new windfarms on land, other developers are attracted to the still comparatively undeveloped open sea.

8.3.5   Offshore Windfarms

With offshore windfarms the turbines are sited directly in the water. For maximum effectiveness the depth of the water should not be too great, and the turbines should not be too far from the coast. Along with the enormous size of the available areas, offshore use also provides other benefits: the wind blows stronger and more evenly on open seas than on land. The power output per wind turbine increases at offshore sites and can be more than 50% higher than at onshore sites.

Offshore wind turbines differ relatively little from onshore ones. Generally, offshore turbines do not require a good deal of maintenance. Access to wind turbines on high seas is not possible in bad weather or when the sea is rough. Special ships are required for major maintenance work but can only be used if the water is relatively calm. Because offshore wind turbines stand in salt water, all components have to be well protected and corrosion-resistant.

Special ships with cranes erect the wind turbines out at sea (Figure 8.15). Special foundations are used to anchor the wind turbines to the seabed. With monopile foundations, a large steel pipe is driven many metres into the seabed. The stability of the bottom is important, and a more secure foundation is needed if the ground is too soft, which increases the cost. This problem may make it uneconomic to erect projects in shallow water.

Figure 8.15 The Nysted offshore windfarm in the Baltic Sea off Denmark. Source: nystedhavmoellepark, http://uk.nystedhavmoellepark.dk. Left: Construction of the windfarm. Photo: Gunnar Britse.

Connecting offshore windfarms to the grid is also more difficult than on land. Sea cables link the different wind turbines to a transformer station, which looks like a small drilling platform. The transformer station converts the electric voltage of the wind turbines to high voltage to keep transmission losses to a minimum. Direct current transmission may be necessary for installations that are far from shore, as losses with alternating-current sea cables can be considerable. In this case, special converters transform the alternating voltage into direct voltage at sea and back into alternating current at onshore substations. Normal high-voltage lines then transport the electricity to the users.

ⓘ Offshore, Onshore and Nearshore

The term *offshore* refers to wind turbines that are sited in the open seas.

Use of the term *onshore* in respect of wind power is relatively recent and only became popular in tandem with the growth of offshore windfarms. *Onshore* means *on land* or *on the mainland*. In this sense onshore windfarms are all windfarms that are not standing in water. These simply used to be called windfarms.

As part of a test of offshore windfarms, some wind turbines were recently erected in the water just a few metres from the shore. Technically, these test facilities were really onshore wind turbines 'with wet feet'. The term *nearshore wind turbines* has also been used to differentiate such installations.

Several large offshore windfarms have been erected in recent years. Denmark has been one of the pioneers, followed by Britain. By contrast, Germany has been rather slow in moving ahead in this field. In part this is due to the major technical challenges involved. German offshore windfarms are planned for comparatively deep waters of 20–50 m at a distance of 30–100 km from the coast. In the meantime, however, a good number of offshore windfarms have also been built in Germany, and others are in the pipeline (see Figure 8.16).

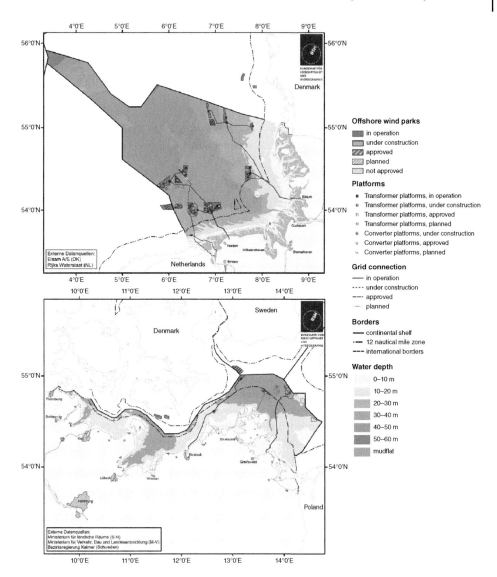

Figure 8.16 Planning for offshore windfarms in the North Sea and Baltic Sea off Germany. Source: Federal Maritime and Hydrographic Agency of Germany, www.bsh.de, as of October 2017.

From a legal point of view, coastal waters are divided into two areas. Territorial waters extend to 12 nautical miles (22.2 km) from shore. The exclusive economic zone (EEZ) then begins and stretches to a maximum of 200 nautical miles (370.4 km). In the Baltic Sea, the German EEZ is much smaller than in the North Sea due to the neighbouring states.

Within Germany's territorial waters, the respective federal states are responsible for the approval of windfarms. Due to the negative effects on the coastline, only very few windfarms are planned in this area. In the EEZ, on the other hand, windfarms are practically invisible from land due to the large distance from the coast. The Federal Maritime and Hydrographic Agency (BSH) is responsible for the approval of windfarms in the EEZ. The BSH assesses whether a planned windfarm obstructs shipping or endangers the marine environment. In the meantime, approval has already been obtained for a large number of projects. An up-to-date overview can be found on the website of the BSH or the IWR.

• www.offshore-windindustrie.de	Offshore wind information by IWR
• www.bsh.de	Federal Maritime and Hydrographic Agency

The construction of offshore windfarms in Germany is not undisputed. For a long time, offshore windfarms in Germany were unable to fulfil hopes for rapid construction of large capacities. Problems and risks in windfarm construction, grid connection and financing have led to long delays in many projects. The remuneration for offshore wind projects has therefore already been increased several times in the past. However, recent tenders for new offshore wind turbines have seen significant price reductions. This suggests that in the medium-term offshore wind power will produce electricity at similar costs to onshore wind power.

The investment costs for large offshore windfarms quickly reach the order of a billion euros. Therefore, only larger energy suppliers are able to handle this kind of project. Another disadvantage of offshore wind energy is the concentration of power generation on the coast, while the large electricity consumers are located in the centre or south of Germany, necessitating numerous new high-voltage transmission lines. In principle, offshore wind power could be converted into renewable gas directly on the coast and then transported without new power lines. Offshore windfarms affect far fewer people than onshore windfarms, which is why the acceptance of offshore wind power is significantly higher. Since the locations for onshore wind power plants and photovoltaic plants in Germany will not be sufficient for the energy transition, a complete decarbonization of the energy supply will not succeed without the use of offshore wind power. For this reason, wind power capacity on land and on the high seas must be rapidly expanded for effective climate protection.

8.4 Planning and Design

Anyone who wants to erect a very small wind turbine in their garden must take into account local planning laws. Overall heights of up to 10 m are usually not a problem. It is also important to ensure that no neighbours will suffer from noise or overshadowing.

Obtaining approval for large windfarms is considerably more difficult. Windfarms can usually only be erected in specifically designated areas identified within the regional planning specifications of local authorities and municipalities. Planning and approval regulations vary depending on the local authority. The local planning department checks the basic feasibility of a windfarm based on a preliminary planning application. If everything is in order, the applicant can submit an official application for the project.

Applications with the relevant forms and explanatory notes can be obtained from the responsible local authority. In addition to checking the safety of the turbines and the location, the relevant authority looks into environmental compatibility and the effect of noise pollution and overshadowing.

Anyone embarking on the sometimes obstacle-strewn path of submitting an application should first ensure that the proposed location is technically and economically suitable for building a windfarm.

The prerequisite for optimal planning is having precise knowledge about the wind conditions at a chosen site, because even the smallest fluctuation in wind supply can have a considerable impact on output. If no measurement data on wind velocity are available for the immediate vicinity of the planned site, it is highly advisable that a measurement station be set up. This station will chart the wind speeds over at least one year. These results should then be compared to the long-term measurement data of other stations and, if necessary, corrected to provide a yearly average. In the case of commercial windfarms, reliable experts should be called in to carry out measurements and calculations and submit a wind survey report.

If usable information about the wind speed is available, computer models can be used to convert the calculations for wind conditions at nearby sites or for other heights. Data on shadowing caused by other wind turbines can also be incorporated into these calculations. The calculations should include the frequency distribution of the wind speed at hub height.

Annual Energy Yield of a Wind Turbine

The frequency distribution $f(v)$ indicates how often a wind velocity v occurs during the year. This usually involves compiling wind velocities at intervals of $1\,\mathrm{m\,s^{-1}}$. The power curve of a wind turbine indicates the electrical output $P(v)$ of the wind turbine at a wind speed v. The power curve is usually provided by the turbine manufacturer. The annual energy output E of a wind turbine is calculated on the basis of both characteristics: the corresponding value of the frequency distribution for each wind speed is multiplied by the performance characteristic and then all the results added:

$$E = \sum_i f(v_i) \cdot P(v_i) \cdot 8760 \ \mathrm{h} = \left(f\left(0\tfrac{\mathrm{m}}{\mathrm{s}}\right) \cdot P\left(0\tfrac{\mathrm{m}}{\mathrm{s}}\right) + f\left(1\tfrac{\mathrm{m}}{\mathrm{s}}\right) \cdot P\left(1\tfrac{\mathrm{m}}{\mathrm{s}}\right) + \cdots \right) \cdot 8760 \ \mathrm{h}$$

The annual energy output for the characteristics from Figure 8.17 is

$$E = (0\% \cdot 0\mathrm{kW} + 4.3\% \cdot 0\,\mathrm{kW} + 8\% \cdot 7.5\,\mathrm{kW} + 10.8\% \cdot 62.5\,\mathrm{kW} + 12.3\% \cdot 205\,\mathrm{kW}$$
$$+ \ 12.6\% \cdot 435\,\mathrm{kW} + 11.9\% \cdot 803\,\mathrm{kW} + 10.5\% \cdot 1330\,\mathrm{kW} + 8.6\% \cdot 2038\,\mathrm{kW}$$
$$+ \ \cdots) \cdot 8760\,\mathrm{h}$$
$$= 11\ 685\ 000\ \mathrm{kWh}$$

With windfarms, deductions are made for factors such as mutual interference (farm efficiency), turbine breakdowns and downtimes (availability) as well as a provision for other unanticipated losses. Various applications available online offer approximate calculations of turbine yield.

Professional planning consultants use sophisticated computer programs for more detailed calculations. However, the principle of the calculations is always the same as

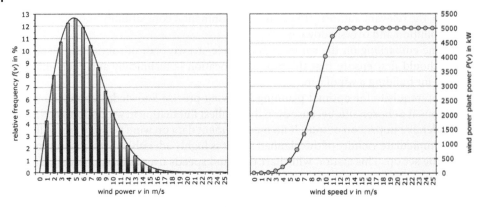

Figure 8.17 Left: Wind speed frequency distribution. Right: Power curve of a wind turbine.

that presented here. In addition to providing calculations on yield, the object of planning is to provide calculations on windfarm optimization. Different aspects are considered. Building higher towers and installing larger rotors will increase the output of an installation. However, this will also increase costs dramatically. The selection of a certain tower height and rotor size will ultimately provide the optimum installation from an economic point of view. Furthermore, losses caused by overshadowing can be minimized through minor changes in the positioning of individual turbines.

- http://www.volker-quaschning.de/ software/windertrag/index_e.html
  Online yield calculations for wind energy converters

8.5 Economics

Very small wind generators are available for a few hundred euros. A simple 500 W wind generator for charging batteries costs about €300. This price does not include the mast and installation. This means that the costs of small wind generators exceed those of photovoltaic systems. Therefore, small wind systems only make economic sense at very windy locations and only then when electricity does not also have to be fed into a grid.

The costs per watt drop as the wind generator size increases. Around €900 kW^{-1}, thus around 90 cents W^{-1}, should be assumed for a grid-connected wind turbine in the megawatt range. The cost of the turbine, tower, and installation is included in this price. The tower and the rotor blades account for around half of the cost (Figure 8.18).

The ancillary investment costs for planning, development, foundations, and grid connections amount to about 30% of the wind turbine costs and need to be added on to the price [DEW02]. Around €1200 kW^{-1} should be assumed for a turnkey windfarm. The project costs for a small farm with four turbines of 2.5 MW each will amount to 12 million euros. Depending on the location and the technology, these costs can vary considerably upwards or downwards. A figure of around 5% of the wind turbine costs should be assumed for annual operation and maintenance.

Windfarms are usually built by operating companies. A group of private investors covers around 20–30% of the cost of erecting the windfarm. The remaining financing

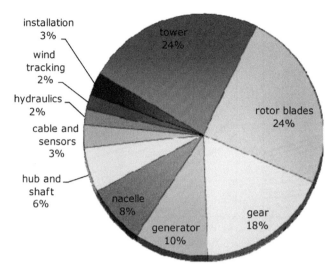

Figure 8.18 Distribution of costs for a 1.2 MW wind turbine [BWE07].

is obtained through bank loans. The feed-in compensation is used for repayment and redemption costs. The surplus funds are distributed among the investors. If everything goes according to plan, private investors will receive a return of 6–10%. If for whatever reason a windfarm does not achieve the output yield forecast, the investors can end up losing all the money they put into the project.

Offshore windfarms cost about twice as much as those on land. The prerequisite for these windfarms is a minimum size of several hundred megawatts. Depending on the distance from shore and water depth, costs can vary considerably between different farms.

In Germany the feed-in-tariffs for wind turbines were determined for many years by the Renewable Energy Sources Act (EEG). Depending on the location, the remuneration varied between 5 and just under 9 cents kWh^{-1} up to 2016. Compensation for feed-in electricity is granted over a 20-year period. Modern wind turbines are therefore designed for a service life of 20 years. For offshore installations the payments were analogous, except that higher rates applied with an initial order of up to 19 cents kWh^{-1}. Although the EEG set a target value for the total annual construction, there was no real volume limitation.

In 2017 the funding procedure, which had been successful for many years, was replaced by a tendering system with fixed payments. The number of installations that can be subsidized is capped. Certain quantity quotas are regularly auctioned off every few months. Only the bidders with the lowest costs are awarded contracts and are allowed to build. Tender volumes in 2017 were far too low to achieve effective climate protection.

In the first tenders, the previous fixed EEG payment rates were clearly undercut. Some people take a very critical view of the extremely low payments offered. Some providers bet on sharply falling system prices. The payments may therefore not be sufficient to finance high-quality installations, which will cause problems in the long term. Furthermore, the low payments mean that there is no longer any scope for financing measures

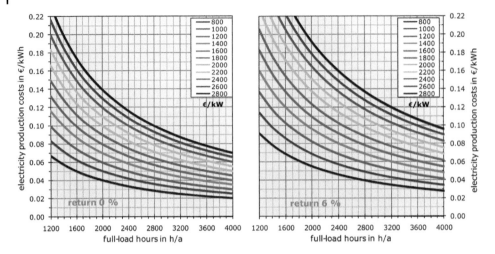

Figure 8.19 Electricity generation costs for wind turbines based on full-load hours and investment costs in €/kW. Assumption: Annual operating and maintenance cost proportion is 5% of the investment cost.

to promote acceptance in the construction of windfarms. This could seriously damage the acceptance of the necessary wind energy expansion in the medium term and thus make the energy transition more difficult or even impossible.

The profitability of a site can be gauged by the number of full-load hours. A good onshore site achieves about 2000 full-load hours per year. In other words, a wind turbine generates as much electricity at the site during the course of a year as if it were operating 2000 hours continuously at full load.

Wind turbines at poor onshore sites reach fewer than 2000 full-load hours. In mountainous areas, on the coast, or with very high wind masts and special low wind turbines, significantly better values can also be achieved onshore. Offshore wind turbines can reach 3000–5000 full-load hours.

Figure 8.19 shows electricity generation costs for wind turbines based on different full-load hours. The assumption for all calculations is that the yearly operating and maintenance costs amount to 5% of the investment costs.

Consequently, on the basis of an inland site reaching 2000 full-load hours and with investment costs of €1200 kW^{-1}, the generating costs amount to 8 cents kWh^{-1} (€0.08 kWh^{-1}) – resulting in an overall return of 6%. Without a return assumption, generation costs fall to €0.06 kWh^{-1}. For an offshore windfarm with 4000 full-load hours per year and investment costs of €3500 kW^{-1}, the generating costs are around €0.12 kWh^{-1} with an expected return of 6%. In the future, a reduction in electricity generation costs is expected due to lower investment and operating costs, a further increase in full-load hours and a reduction in risks.

8.6 Ecology

Due to their size, wind turbines are often visible even from great distances and therefore stand out in many places. Whether one finds wind turbines attractive or ugly is really a matter of personal taste. Nevertheless, the subject has become a major bone of

Figure 8.20 Wind turbines in rural landscapes. Source (left): German Wind Energy Association. Source (right): REpower Systems AG, Photo: Jan Oelker.

contention. Whereas supporters swoon over the majestic appearance of these technical masterpieces, opponents fight against what they consider a blot on the landscape. There certainly are arguments that can legitimately be made against wind turbines. However, it is important to bear in mind that wind turbines are usually erected in areas that already bear many signs of human activity.

Traditional windmills have been a common sight for many centuries, and are now regarded as picturesque additions to the landscape. Domestic livestock seems to get used to the turbines much more quickly than humans (Figure 8.20). It is incredible to think that around 190 000 high-voltage pylons and more than one million medium and low-voltage masts are less controversial in Germany than the 30 000 or so wind turbines. The strong feelings about wind turbines may arise because people are less used to seeing them than the other power infrastructure that criss-crosses the European landscape.

Considered objectively, the negative effects are limited. Erecting wind turbines in conservation areas is just as much a taboo as it would be in residential areas. If wind turbines adhere to a minimum distance from residential areas, the nuisance from noise or overshadowing will be minor. These issues are part of the approval process in the planning of windfarms.

Individual federal states are making the expansion of wind power considerably more difficult with excessive distance regulations. In Bavaria, the so-called 10-H rule was adopted in 2014. According to this, wind turbines must maintain a distance of 10 times the height of the wind turbine from buildings, which means that only very few areas remain suitable for the construction of wind turbines. As a consequence, more electricity must be generated by offshore wind turbines, which in turn results in a much greater demand for power lines.

Wind power has very little effect on the animal world. Most animals get used to the turbines quickly. Birds usually detect the relatively slow-rotating rotors from a great distance and fly around them. Despite the claims made by the opponents of windfarms, birds very seldom fly into a turbine. The numerous window panes in residential buildings pose a far greater threat to the bird world than wind turbines do.

The climate protection balance of wind power is very positive. Compared to photovoltaics, the energy required to manufacture wind turbines is lower. A wind turbine will recoup this amount within a few months. If a wind turbine displaces electricity from coal, it avoids emissions of around 1000 g of carbon dioxide per kilowatt hour fed into the grid. A single 5 MW wind turbine at a good location will generate around 200 million kilowatt hours of electrical energy in 20 years and can thus avoid 200 000 t of carbon dioxide.

8.7 Wind Power Markets

Denmark is rightfully considered the pioneer of modern wind power. The Danish teacher Paul la Cour built a wind turbine for generating electricity at the primary school in Askow as far back as 1891. When the current wind power boom began in the 1980s, Danish companies ranked among the leaders of this technology from the beginning. German companies were also involved right from the start. Today, the market leader in Germany is Enercon with around 13 000 employees worldwide. In recent years, China has taken an impressive leadership role. Due to the gigantic number of installations in their own country, the Chinese manufacturers now make up the largest wind companies worldwide. In 2015, the largest Chinese company, Goldwind, installed significantly more wind turbines in China alone than all manufacturers in the whole of Germany.

The USA experienced its first wind power boom at the end of the 1980s, by which time it had already erected thousands of wind turbines in just a few years. However, the wind power market in the USA came to an almost complete halt in the early 1990s. Due to a legally regulated system of feed-in-tariffs, Germany then became number one in the wind energy market in the 1990s and was able to retain this position until 2007. As a result, Germany developed an internationally successful wind energy industry, with a turnover in the billions and a high export ratio (see Figure 8.21).

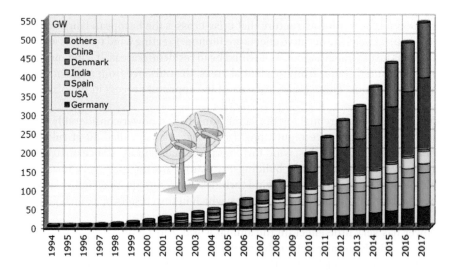

Figure 8.21 Development of wind power capacity installed worldwide.

Besides Germany, China, the USA, Spain and India are among the most important wind power markets today. The wind energy market in the USA is driven by tax write-offs and quotas for renewable energies. Due to changing conditions in the past few years, the market there has been very volatile. India has also developed a steady wind energy market and, consequently, a flourishing wind energy industry. The Indian Suzlon Group is one of the biggest wind turbine manufacturers in the world.

Since 2009, China has been the leader in installed wind power capacity and is continuously expanding its lead. In 2014 and 2016 alone, China installed more wind turbines than Germany had between 1990 and 2016. Numerous other countries have also started to expand their use of wind power in recent years, so that high annual growth rates can still be expected in the wind industry. Current figures on market developments can also be found on the Internet.

- www.wind-energie.de   German Wind Energy Association
- www.ewea.org   European Wind Energy Association
- http://www.gwec.net   Global Wind Energy Council

8.8   Outlook and Development Potential

Whereas only very few countries were players in the wind boom of the 1990s, in the interim many more have started to rely on wind energy. Even today at good sites wind energy is able to compete with conventional fossil power plants. In contrast with the fluctuating prices for coal, natural gas and crude oil, the production costs for a wind turbine are quite constant, once it has been erected. It is therefore anticipated that the high growth rates for wind power will be sustained worldwide.

In Germany, the market stagnated between 2002 and 2012 (Figure 8.22). Development accelerated from 2013. This was driven above all by the government's announcements

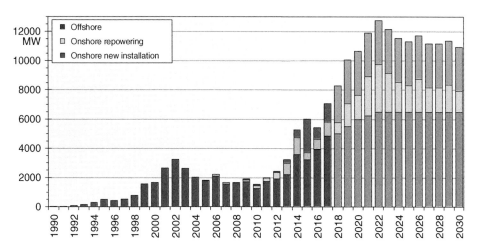

Figure 8.22 Development of newly installed wind power capacity in Germany until 2017 and necessary expansion for a climate-compatible energy transition until 2030.

to lower the expansion targets and to create new, less favourable framework conditions for wind energy expansion through tenders.

For a rapid energy transition and effective climate protection, however, the expansion of wind energy would have to be stepped up further. According to a study by the German Wind Energy Association, up to 2% of the land area in Germany could be used for the installation of wind turbines [BWE11]. Theoretically, up to 200 000 MW of wind power capacity could be built. This kind of figure is required for the complete decarbonization of Germany. At the end of 2017, the installed capacity was only just under 51 000 MW.

As wind turbines have a service life of about 20 years, Germany is successively developing a new market segment: repowering. With repowering, new, more efficient installations replace older and no longer viable smaller wind turbines. As a result, the installable onshore wind capacity is increased.

The development of the German offshore locations has been very slow for many years. Installation figures did not start to rise noticeably until 2014. For a climate-neutral energy supply, the expansion of offshore wind energy must also be further accelerated.

Overall, wind power could cover between 40% and 50% of Germany's electricity demand by 2030 and become one of the most important pillars of a climate-compatible electricity industry. Since other countries are following Germany's example, and China has already clearly overtaken Germany in terms of installation figures, wind energy could well meet more than a quarter of global electricity demand by 2050.

9

Hydropower Plants – Wet Electricity

Today there are far fewer hydroelectric plants than there were in their heyday at the end of the eighteenth century. At that time between 500 000 and 600 000 water mills were running in Europe alone [Köni99]. In those days France was the country with the highest number, although thousands were also being used in other parts of Europe. Water wheels not only drive mills but also power a number of other tools and machines. Flowing waterfalls were harnessed by watermills with wheels up to 18 m in diameter. At five to seven horsepower the average power of water mills at that time was still comparatively modest (Figure 9.1).

As more and more mills were built on rivers and streams, the activity was strictly regulated, and mill operators were instructed on how long mills could be used and how large they could be. This may have been an annoyance for them, but it was a good thing because it promoted technical development and ensured that optimal use was made of existing mills. This constant drive to improve gave rise to the modern, highly efficient turbines used in hydropower plants today.

The introduction of the steam engine slowly displaced water-powered systems. But, in contrast to wind power, the use of hydropower did not vanish with the increased exploitation of fossil energies. When widespread electrification began at the end of the nineteenth century, hydropower was still very much part of the scene. At the beginning small turbines were used to power electric generators, but the size of the systems grew rapidly.

9.1 Tapping into the Water Cycle

The colour of our planet is blue when seen from space. The reason is that 71% of the Earth's surface is covered by water. However, without the sun our blue planet would not be blue. Water, which gives the Earth its characteristic appearance, would be completely solidified into ice, as 98% of water is liquid purely due to the sun's heat.

A Water-Powered System in the Rain Gutter?

The roof of a house collects many cubic metres of water each year. A rain gutter diverts the water without making any use of its energy. A water powered system for each rain gutter could actually be a good idea.

Renewable Energy and Climate Change, Second Edition. Volker Quaschning.
© 2020 John Wiley & Sons Ltd. Published 2020 by John Wiley & Sons Ltd.

Figure 9.1 Historic watermill in the Alps. Source: Verbund, www.verbund.at.

The annual precipitation in Berlin amounts to around 600 l per square metre, so that a 100 m² house roof would collect 60 000 l or 6000 10-l buckets of water. Compared with a similar bucket on the ground, the content of a 10-l bucket on a 10 m-high roof has a potential energy of 0.000273 kW h. Six thousand buckets together create around 1.6 kW h, just enough to boil water for 80 cups of coffee, but unfortunately too little to make it worthwhile to harness. Thus, considerably larger quantities of water than are available in a rain gutter are necessary to produce viable quantities of energy.

Altogether there are around 1.4 billion cubic kilometres of water on Earth. Of this amount, 97.5% is salt water in the oceans and only 2.6% is fresh water. Almost three-quarters of the fresh water is bound in polar ice, ocean ice, and glaciers; the rest is mainly in the groundwater and soil moisture. Only 0.02% of the water on Earth is in rivers and lakes.

Due to the influence of the sun, on average 980 l of water evaporate from each square metre of the Earth's surface and return elsewhere in the form of precipitation. Altogether about 500 000 cubic kilometres of water collect in this way each year. This gigantic water cycle converts around 22% of all the solar energy radiated onto Earth (Figure 9.2).

If the evaporation were concentrated onto a single square kilometre on Earth, the water would come pouring out at a speed of over 50 km h^{-1}. The column of water would reach the moon in less than one year. Fortunately, however, the water remains spread on the surface of the Earth, because otherwise we would not have any liquid water left in 3000 years.

The energy that is converted by the sun in the Earth's water cycle is around 3000 times the primary energy demand on Earth. If we were to stretch a tarpaulin at a height of 100 m around the Earth, collect all the rainwater into it and use it to produce energy, we would be able to satisfy all the energy needs on Earth.

Altogether 80% of precipitation comes down over the ocean, while a large amount of the 20% falling on land evaporates. Around 44 000 cubic kilometres of water reach the

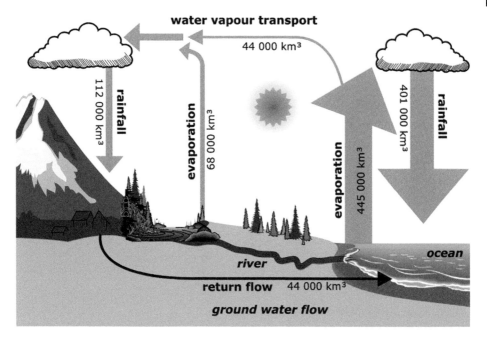

Figure 9.2 Earth's water cycle.

ocean again as return flow in groundwater or in rivers. This is still more than one billion litres per second. We could be making use of the energy of the water from this return flow without stretching large tarpaulins over the Earth. This water could be delivering some of its energy on its way back to the ocean – energy that it first received from the sun. Yet the usable amount of energy that rivers carry with them comprises only a small fraction of the energy of the water cycle.

To convert water into power, it is not just the amount of water that is important but the height at which water is found. Water in a small mountain stream with many hundreds of metres of difference in elevation can sometimes carry more energy than a large river that only has to descend from an elevation of a few metres to reach the ocean. About a quarter of the energy of water from rivers and lakes can theoretically be used. This equates to around 10% of current global primary energy demand. Ocean currents and waves also contain usable energy.

9.2   Water Turbines

Water turbines form the core of water-powered systems and extract the energy from water. Modern water turbines have very little in common with the rotating wheels of traditional watermills. Depending on the head of the water and the water flow, turbines optimized for the respective operating area are used (Figure 9.3). These turbines reach a power of over 700 MW.

The Kaplan turbine, developed by the Austrian engineer Viktor Kaplan in 1912, is usually the first choice for low heads – for instance, for power plants on rivers (Figure 9.4). This turbine uses the pressure of the water at the different elevations of barrages. It has three to eight adjustable blades and looks like a large ship screw (propeller) that

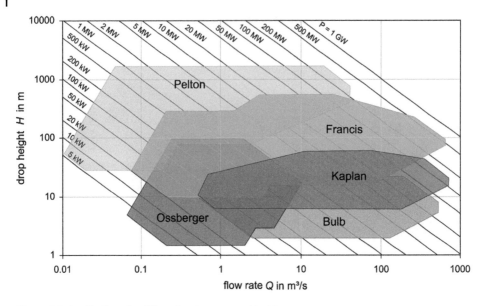

Figure 9.3 Applications for different water-powered turbines.

powers the flowing water. The efficiency of Kaplan turbines reaches values between 80% and 95%.

The tubular turbine (Figure 9.5) is similar to the Kaplan turbine, except that it has a horizontal axis and, as a result, is suitable for even smaller heads. The generator is placed in a bulb-shaped workroom behind the turbine, which explains why it is also referred to as a bulb turbine.

Figure 9.4 Drawing showing a Kaplan turbine with a generator (left) and a photo of a Kaplan turbine (right). Source: Voith Hydro.

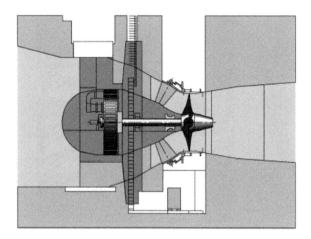

Figure 9.5 Bulb turbine with generator. Source: Voith Hydro.

For larger heads of up to 700 m the Francis turbine is used. It was developed in 1848 by James Bicheno Francis, who was born in England and emigrated to the United States when he turned 18. This turbine also uses the pressure difference of water and reaches efficiencies of over 90%. In principle, the Francis turbine can also be used as a pump and is therefore suitable as a pump turbine for pumped-storage plants (see Figure 9.6).

In 1880 the American Lester Allen Pelton developed the Pelton turbine. It is mainly designed for large heads and, consequently, for use in high mountains. This turbine can reach very high efficiencies of 90–95%. The water is supplied to the turbine over a penstock. It then flows through a nozzle at very high velocity to spoon-shaped buckets.

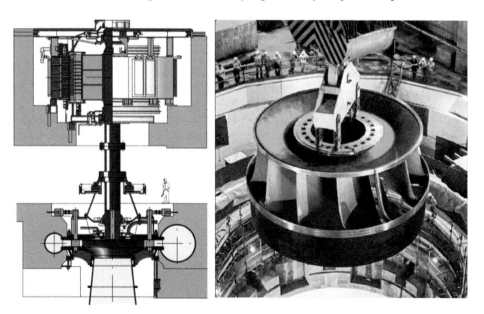

Figure 9.6 Francis pump turbine at Goldisthal pumped-storage plant (left) and a Francis turbine at the Itaipu plant (right). Source: Voith Hydro.

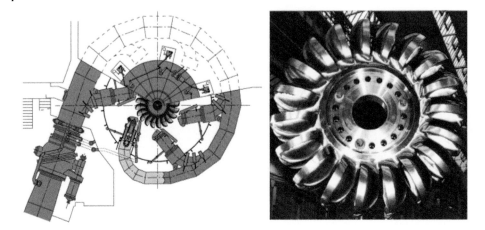

Figure 9.7 Drawing of a six-nozzle Pelton turbine (left) and photo of a Pelton turbine (right). Source: Voith Hydro.

Small plants use the Ossberger turbine, which is also called a cross-flow turbine. These turbines reach somewhat lower efficiencies of around 80%. The turbine is divided into three parts that can be powered separately by water. This helps the system cope with fluctuations in the water runoff of small rivers (see Figure 9.7).

9.3 Hydropower Plants

The energy that can be exploited from water essentially depends on two parameters: runoff volume and the head of the water. Almost all hydropower plants utilize natural elevation differences using technical equipment.

9.3.1 Run-of-River Hydropower Plants

The natural course of a river itself concentrates large quantities of water. A run-of-river hydropower plant can be built anywhere on a river where a sufficient difference in elevation exists (Figure 9.8). A weir then backs up the water. This creates a difference in elevation in the water's surface above and below the plant. At the top the water flows through a turbine that powers an electric generator. Grating at the turbine intake

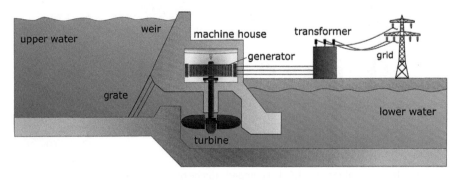

Figure 9.8 Principle of a run-of-river hydropower plant.

prevents rubbish and flotsam washed along by the river from blocking up the turbine. A transformer then converts the voltage of the generator into the desired mains voltage.

Large hydropower plants are usually constructed so that multiple turbines can run in parallel. If the water flow drops during dry periods of the year, some of the turbines can be shut down. The remaining turbines then still receive almost the full amount of water they need. This prevents the turbines from working in partial-load mode with poor efficiency. If, on the other hand, there is flooding, and a river carries more water with it than the turbines can process, the surplus water must be let out, unused, over a weir.

Barrages can present an obstacle for shipping and animals that live in rivers. Locks installed parallel to a barrage enable ships or boats to overcome the difference in water height. A certain amount of residual water flows through a fish bypass system next to the weir. This enables fish and other aquatic creatures to move freely across the weir system.

Germany's largest run-of-river power plants are located on the Rhine. Many plants, such as the run-of-river plant at Laufenberg on the Rhine (Figure 9.9), were built in the first half of the twentieth century. Existing plants that have been modernized have been able to improve performance. The best example in Germany is the Rheinfelden power plant, which was built in 1898 and is currently being modernized. The new facility, due for completion in 2011, will increase power from 25.7 to 100 MW and more than triple electricity production. At the same time modern fish bypasses are improving the ecological situation.

Europe has very strict environmental conditions for new energy plants and for those being modernized. Consequently, it is not anticipated that electricity generation from run-of-river power plants will increase substantially.

As the height difference in water levels with run-of-river plants is usually only a few metres, very few plants have an output of more than 100 MW of electric power. Furthermore, run-of-river plants tend to be difficult to regulate. They supply electricity

Figure 9.9 Run-of-river hydropower plant at Laufenburg. Source: Energiedienst AG.

Figure 9.10 Examples of storage power plants in Austria: Malta (left), Kaprun (right). Source: Verbund, www.verbund.at.

around the clock, but because the currents of a river cannot be controlled, surplus water remains unused if the power is not needed.

9.3.2 Storage Power Plants

Storage or impoundment hydropower plants produce high levels of power output. Dams store huge masses of water at geographically optimal locations. This type of dam makes it possible for storage power plants to be built in the mountains (Figure 9.10). A high-pressure pipeline pumps the water into a machine house, where enormous water pressure of up to 200 bars is created. In the machine house, water powers the turbines that produce energy via an electric generator.

It is not unusual to find dams over 100 m high. The tallest dams on Earth are over 300 m high. Reservoirs are often also used to store drinking water and to control flooding. The Rappbode Dam in the Harz Mountains is the tallest dam in Germany at 106 m. It is mainly used for storing drinking water. The power plant output is comparatively low at 5.5 MW.

If storage hydropower plants are designed mainly for power generation, they achieve significantly higher outputs. Power plants with several hundred or even thousand megawatts are not uncommon (Table 9.1).

Storage power plants can be constructed with a second basin located at a lower elevation by the machine house. Special pump turbines can then pump the water from the lower basin back up to the higher basin. This is called a pumped-storage hydropower plant.

9.3.3 Pumped-storage Hydropower Plants

Pumped-storage power plants need very specific geographical conditions. To be viable, they need two basins with a significant difference in altitude between them. Some pumped-storage plants have a natural inflow where a river flows into the higher basin. Pumped-storage plants without a natural inflow are purely storage plants.

Table 9.1   The largest hydroelectric power plants in the world

Power plant	Country	River	Completed	Capacity in MW	Dam length in m	Dam height in m
Three Gorges	China	Yangtse	2008	22 400	2335	181
Itaipú	Brazil/Paraguay	Paraná	1983	14 000	7760	196
Xiluodu	China	Jinsha Jiang	2014	13 860	698	286
Guri	Venezuela	Rio Caroni	1986	8 850	7500	162
Tucuruí	Brazil	Rio Tocantins	1984	8 370	6900	106
Grand Coulee	USA	Columbia River	1942	6 809	1592	168

When electric power is needed, the water of the higher basin is pumped over an intake mechanism through a pressure pipe to the turbine, which drives an electric machine as the generator. Once the turbine has extracted the energy from the water, the water flows into the lower basin. There, a transformer converts the voltage of the generator to mains voltage (Figure 9.11).

When electricity surpluses occur in the grid, the pumped-storage plant switches over to pumping operation. The electric machine then functions as a motor that drives the pump turbine. The pump turbine pumps the water from the lower to the higher basin once again. Major fluctuations in pressure can occur when the generator is switched to pumping operation, and, in extreme cases, these fluctuations can cause damage to the pressure pipe or other system parts. A surge tank regulates this change in pressure.

Pumped-storage power plants reach efficiencies of 70–90%. A good 70% of the electric energy that is necessary for pumping the water to the higher level can be replaced during the operation of the generator. Despite the losses, pumped-storage plants are very attractive economically. When a surplus of supply exists, electricity can usually be obtained very inexpensively. On the other hand, if electricity is in low supply, a plant can feed the energy back into the grid at a higher price.

Pumped-storage power plants have grown in importance in recent years. They can help to compensate for fluctuations from renewable power plants such as photovoltaic or wind power plants. For a fully renewable electricity supply, however, a considerably

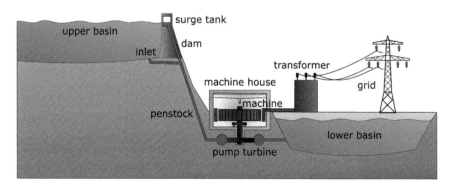

Figure 9.11 Principle of a pumped-storage power plant.

Figure 9.12 The Goldisthal pumped-storage power plant in Germany. Source: Vattenfall Europe.

larger storage capacity is required than pumped-storage power plants in Germany could supply.

Germany's largest pumped storage power plant is located in Thuringia. The Goldisthal power plant (Figure 9.12), commissioned in 2003, has four generators which together have an output of 1060 MW. The accumulated dam volume of the upper basin is 12 million cubic metres. Based on this volume and with an average head of 302 m, the power plant can work at full capacity for eight hours. This is sufficient to cover the electricity demand of over 2.7 million average households.

9.3.4 Tidal Power Plants

Tidal waves are attributed to the interaction of the forces of attraction between the moon, the sun and Earth. As a result of the rotation of the Earth, the forces of attraction are continuously changing their direction. The water masses of the oceans follow the attraction. As a result, a tidal wave with a height difference of more than 1 m can form on the open sea. Tidal waves caused by the moon occur approximately every 12 hours at some point on Earth. In extreme cases, they can reach more than 10 m in height. Tidal power plants could use these changes in water level to generate energy.

In regions with a high tidal hub, a reservoir can be built so that water flows into it at high tide. The incoming water flows into the reservoir through a turbine in the dam wall, and at low tide it flows back again. The turbine and the connected generator convert the energy of the water into electric energy. The energy output is not continuous, so output drops to zero when the tide goes out.

Tides were used by tidal mills even during the Middle Ages. Worldwide there are only very few modern tidal power plants in operation today. The best-known one is the Rance power plant in France, which is situated on the estuary of the river of the same name and came into operation in 1967. It has a power output of 240 MW. The dam wall that seals off the 22 km² basin is 750 m long. The environmental impacts of the separation of the estuary from the ocean are considerable. Corrosion by salty sea water is also causing some problems.

Tidal power plants are comparatively expensive to operate, so it is not likely that many new plants will be built. Due to the relatively minor differences between ebb and flow, most countries do not have suitable locations for these plants.

9.3.5 Wave Power Plants

For decades, high hopes have been pinned on the development of wave power plants. A study of the potential of wave energy indicates that large amounts of energy could be generated. However, only coastal regions with low water depths are appropriate for the use of wave energy. Due to the comparatively small usable sea areas available in many countries, this technology has relatively low potential.

Basically, a distinction is made between the following functions:

- Float systems
- Chamber systems
- TapChan systems

Float systems use the energy potential of waves. A floating body follows the movement of waves. Part of the installation is anchored to the ground. The movement of the floating body can be used by a piston or a turbine.

With chamber systems, a chamber locks in air. The waves cause the water level to fluctuate in the chamber. The oscillating water level compresses the air. The displaced air escapes through an opening and powers a turbine and a generator. When the water column falls, the air flows back through the turbine into the chamber (Figure 9.13).

TapChan is an abbreviation for 'tapered channel'. In these systems, waves in coastal areas or on a floating object feed into a tapered and rising channel. An upper basin captures the waves. When the water flows back into the ocean, it powers a turbine.

Although numerous prototypes have been developed for wave power plants in recent decades, they have not yet caught on to any great extent. The main problem is the

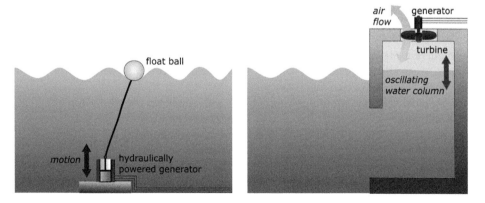

Figure 9.13 Principle of wave power plants. Left: Float system. Right: Chamber system.

Figure 9.14 Left: Prototype system in the Seaflow project off the west coast of England. Photo: ISET. Right: Maintenance ship in a planned ocean current power plant park. Graphics: MCT.

extreme fluctuation in conditions at sea. On the one hand, technical systems must be designed to save on materials and be cost-effective; on the other hand, storms with giant waves place extreme pressure on system survival. Many prototype installations have already fallen victim to storms. However, large companies have now become involved in the development of wave power plants, so it is likely that these problems will be solved in due course.

9.3.6 Ocean Current Power Plants

Ocean current power plants have a structure similar to wind power plants, except that the rotors rotate below the water surface. A built-in hub mechanism raises the rotor to the water surface for maintenance. The first prototype was successfully installed in 2003 off the North Devon coast in England [Bar04] (see Figure 9.14).

In principle, the physical characteristics of wind power plants are similar to those of ocean current plants. The main difference is that, because water has a higher density than air, ocean current plants can achieve higher output yields than wind power plants even when the speed of currents is low.

Ocean current plants are limited to regions with relatively consistent high current speeds and moderate water depths of up to 25 m. These conditions mainly exist on headlands and bays, between islands and in straits. Although shipping lanes often restrict this kind of use, major potential exists. In Germany, for example, it would make sense to build a plant at the southern tip of Sylt. In the medium term, ocean current power plants could become another building block in the generation of a climate-compatible electricity supply, as technological advances and the benefits of mass production are rapidly leading to cost reductions.

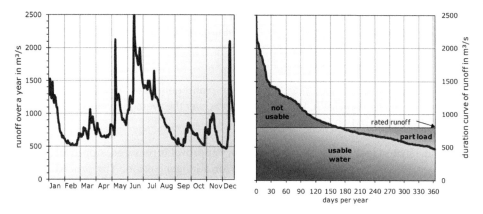

Figure 9.15 Annual flow characteristics and annual continuous curve for the Rhine runoff near Rheinfelden.

9.4   Planning and Design

All types of hydropower plants require different and sometimes complex planning and design. It is only possible here to outline some of the basic planning aspects. The first requirement for a run-of-river power plant is the collection of information about the stretch of water where the plant is to be built. The most important parameter is the runoff of the river's water over the course of a year – that is, the volume of water that flows through the river. Each river has its own typical annual flow characteristic that is influenced mainly by rain and melting snow. For further design purposes, the river runoff volumes are categorized to obtain an annual continuous curve. This indicates the number of days per year when a river achieves or exceeds a certain runoff volume (Figure 9.15).

A rated runoff for a river is established, which represents the water volume at which a power plant produces its full output. If the runoff volume exceeds the rated runoff, the surplus water must be diverted unused over a weir. To maximize electricity generation, turbines with a high rated runoff should be selected. However, if the runoff volume of the river drops below the rated runoff, there will not be enough water to enable the power plant to provide its full output. The turbines will then either work at poor efficiency in partial load mode or individual turbines will be switched off and remain unused. A low rated runoff should be selected to ensure that optimal use is being made of the turbines. In practice, the rated runoff is determined based on a compromise between the maximum amount of power to be generated and optimal use of the turbine.

Power Output of a Run-of-River Power Plant

If the water runoff quantity Q and the head H of the water at a power plant are known, then the electric power output of a hydropower plant can be calculated relatively easily, based on the efficiency η of the plant, the density of the water ρ_W ($\rho_W \approx 1000\ \mathrm{kg\,m^{-3}}$) and the gravitational acceleration g (g $= 9.81\ \mathrm{m\,s^{-2}}$):

$$P_{el} = \eta \cdot \rho_W \cdot g \cdot Q \cdot H$$

With an extension runoff quantity of $1355\,\mathrm{m^3\,s^{-1}}$, a head of 10.1 m and efficiency of $79\% = 0.79$, the output for full load at the Laufenburg hydropower plant on the Rhine can be calculated as

$$P_{el} = 0.79 \cdot 1000\,\mathrm{kg\,m^{-3}} \cdot 9.81\,\mathrm{m\,s^{-2}} \cdot 1355\,\mathrm{m^3\,s^{-1}} \cdot 10.1\,\mathrm{m} = 106\,\mathrm{MW}.$$

At 5940 full-load hours per year, the power plant generates $106\,\mathrm{MW} \times 5940\,\mathrm{h} = 630\,000\,\mathrm{MWh} = 630$ million kWh. This is sufficient to cover the average electricity consumption of 180 000 households in Germany. ■

To build a hydroelectric power plant, statutory approval is required. Approval is usually granted for 30 years, but it may be given for a longer period. The procedure for obtaining approval can take 3–10 years. For large-scale power plants, the approval also includes a test for environmental compatibility. The EU water rights guidelines stipulate that all water bodies in Europe must be kept in a 'good state'. Although the water may be used, its ecological functioning must remain intact. In this context the current state of the water body concerned is also crucial. Therefore, it is much easier to gain approval to modernize a power plant on a stretch of water that has already been greatly altered than to build a new plant. For this reason, the chances of obtaining approval to build new large-scale hydropower plants are not very favourable.

9.5 Economics

Hydroelectric power plants are considered the most cost-effective option for supplying renewable electricity today. This applies mainly to older systems where the construction costs have largely already been written off. The relatively high construction costs and long amortization periods for new sites substantially increase the cost of generating electricity.

For small plants, fewer than five megawatts, the investment in modernization amounts to between €2500 and €5000 kW^{-1}; for the reactivation of a plant or a new plant it is between €5000 and €13 000 kW^{-1}. The costs for larger plants are somewhat lower but depend greatly on local conditions. In addition to the investment and operating costs, large-scale hydropower plants may also have to pay a fee for water use.

With existing medium-sized plants in the power range between 10 and 100 MW, the cost of generating electricity is less than 2 cents per kilowatt hour. For new plants this cost can increase up to 4–10 cents per kilowatt hour [Fic03]. The costs for small plants can even be higher.

In Germany, payments for electricity from new hydropower plants are regulated by the Renewable Energy Sources Act (EEG). Accordingly, the revenue for electricity from hydropower plants with a capacity of less than 500 kW amounts to 12.4 cents per kilowatt hour; plants between 500 kW and 2 MW receive 8.17 cents per kilowatt hour for 20 years. Lower rates are envisaged for larger capacities, so that from 50 MW this is only 3.47 cents per kilowatt hour. From 2018, an annual reduction of 1.5% is also envisaged.

9.6 Ecology

Hydropower plants are among the most controversial of all the renewable power plants. Conventional run-of-river, storage, and tidal power plants, in particular, have a huge impact on the natural environment.

Figure 9.16 Left: Stream in the area of the Donaukraftwerk Freudenau power plant in Austria. Source: Verbund, www.verbund.at. Right: Fish ladder at the Weser power plant in Bremen. Source: Weserkraftwerk Bremen GmbH.

Due to the barrage system, bodies of stationary water form in places where water flowing over pebbles and boulders previously provided a suitable habitat for many different kinds of fish. As a result of the changes to habitat, numerous fish and plants are becoming extinct. Another danger for fish is the hydroelectric turbines themselves. Although grating prevents large fish from entering the processing area of a plant, the metal bars of the grating let small creatures slip through and become injured or killed by the turbines. The barrier systems are often an insurmountable obstacle for migratory fish. Fish ladders that run alongside the barriers help enormously in improving the freedom of movement for fish (Figure 9.16). Nevertheless, barrier systems are still an obstacle for some species.

Large reservoirs flood wide areas of land and destroy people's homes as well as natural habitat. Sinking biomass decomposes in water and releases large quantities of climate-damaging methane. This problem can be reduced considerably if reservoir basins are carefully cleared before they are flooded.

Breaches in walls are another danger with large dams. Normally, dams are constructed so that they are largely earthquake-proof. But even the best construction is powerless against targeted terrorist attacks. Huge devastation will occur if the large volumes of stored water pour into a valley all at once.

The newly built Three Gorges dam in China is often cited as a negative example of the type of environmental damage hydropower plants can cause. Twenty cities and more than ten thousand villages, which together housed more than one million people, were sacrificed for the building of this plant. It is still unclear what the ecological impact has been on the flooded area. It is expected that the many environmental sins of the past will come back to haunt those who live in the area, for example by contaminating the groundwater. The fear is that the 600 km-long reservoir is turning into a cesspit of sewage and industrial waste.

On the other hand, the Three Gorges dam is designed to produce 84 billion kilowatt hours of electricity per year. This equates to more than one-eighth of total German demand. If modern coal-fired plants were used to generate the same amount of electricity, they would be emitting more than 70 million tons of carbon dioxide into the atmosphere each year. This corresponds to the total carbon dioxide emissions of Austria.

In this sense it is important to weigh the benefits and disadvantages of all hydropower plants. It is possible to build ecologically viable plants. The main thing is that ecological aspects are included in the equation along with protecting the environment and providing reasonably priced energy.

9.7 Hydropower Markets

About 16% of the electricity generated worldwide comes from hydropower plants. In 2004 Canada was still the leader in this field (Figure 9.17), but in the meantime it has been overtaken by China (Figure 9.17). The hydroelectric share of the power generated varies substantially from country to country. Whereas 100% of Norway's electricity comes from hydropower plants, this share is around 60% in Brazil, Canada, and Austria. The hydroelectric share for China and the USA is only 19% and 6%, respectively. At 3%, Germany's share is relatively insignificant. In Europe, Norway and Iceland have the highest percentage of hydroelectric energy, followed by the Alpine countries and Sweden.

Large-scale plants, in particular, make a major contribution to the generation of electricity from hydropower. The Itaipu facility in Brazil, with the highest capacity in the world (Figure 9.18), generates more electricity than all hydropower plants in Germany and Austria combined.

In Germany, hydropower is mainly used in the southern federal states, where the rivers have a significantly greater gradient than in the north. In the flat north there are only a few small installations.

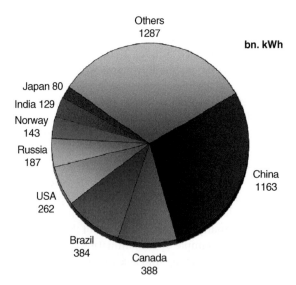

Figure 9.17 Electricity generation from hydropower plants in different countries. 2016. Source: Data: [BP17].

Figure 9.18 Aerial view of the Itaipu power plant. Photo: Itaipu Binacional, www.itaipu.gov.br.

9.8 Outlook and Development Potential

Hydropower is the most developed method for power generation based on renew-ables. In the industrialized nations most of the potential has already been exploited. The potential for new large-scale plants now mainly exists in developing and emerging nations. In many industrialized countries there is also great potential for modernizing or upgrading barrages and dams. Classic run-of-river and storage hydropower plants could at best double the output of electricity generated worldwide. However, in the long term their share of energy production worldwide will decrease, because the demand for electricity will continue to rise.

Certain types of hydropower plants, such as wave and ocean current plants, which are not yet in use, show some promising potential for the future. However, substantial cost reductions are still needed before these types of power plants can have any major impact on the market.

The biggest advantage of hydroelectric energy is its relatively consistent output of power, compared to solar energy and wind power. This aspect makes it easier to plan a mix of different renewable energies. With the increase in renewable energy's con-tribution to electricity generation, storage, and pumped-storage power plants are also becoming important options due to their ability to offer consistency in electricity supply.

10

Geothermal Energy – Power from the Deep

When planet Earth came into being more than four billion years ago, its form was considerably different from what it is now. At that time Earth was in a partially melted state. It was not until about three billion years ago that the temperature of the Earth's surface dropped to below 100 °C and the Earth's crust gradually began to harden.

It may not seem like it in the depths of a northern winter, but today our planet is anything but a cold ball. About 99% of the Earth is hotter than 1000 °C and 90% of the rest has temperatures of over 100 °C. Fortunately for us, these high temperatures are almost exclusively found in the Earth's interior. Every so often volcanoes produce impressive eruptions spewing molten matter from depths of up to 100 km. Different geothermal energy technologies enable us to tap the heat of the Earth's interior in a controlled way so that we can satisfy some of our heat and electricity demands (Figure 10.1).

10.1 Tapping into the Earth's Heat

The Earth itself is made up of concentric bands (Figure 10.2) comprising the core, the mantle, and the crust. The Earth's core has a diameter of around 6900 km (4290 mi). A differentiation is made between the outer, liquid core, and the inner core, which is made of solid matter. Maximum temperatures in the Earth's core reach 6500 °C, which is hotter than the surface of the sun.

Why the Earth's Inner Core is Not Liquid

The Earth's inner core, which essentially consists of iron and nickel, has temperatures of up to 6500 °C. Under normal environmental conditions with an ambient pressure of one bar, these temperatures would make iron and nickel gaseous. The pressure rises as the depth into the Earth's interior increases. This pressure reaches maximum values of four million bars. This extremely high pressure ensures that the outer core is liquid, and the inner core is solid.

Eighty percent of the Earth's core consists of iron. A small amount of the heat in the Earth's interior comes from residual heat from the time the Earth was formed but a major part consists of radioactive decaying processes. The Earth's mantle is around 2900 km thick. We are only able to reach the top part of the Earth's crust.

Renewable Energy and Climate Change, Second Edition. Volker Quaschning.
© 2020 John Wiley & Sons Ltd. Published 2020 by John Wiley & Sons Ltd.

Figure 10.1 Volcanic eruptions bring the energy from the Earth's interior to the surface in a spectacular fashion. Source: D.W. Peterson and R.T. Holcomb, US Geological Survey.

The Earth's crust and the uppermost part of the Earth's mantle together form the lithosphere. The thickness of the lithosphere varies from a few kilometres to more than 100 km. It consists of seven large and some smaller lithosphere plates (Figure 10.3). These rather brittle plates float in the asthenosphere, where matter is no longer solid. The plates are constantly in motion. Earthquakes and volcanoes frequently

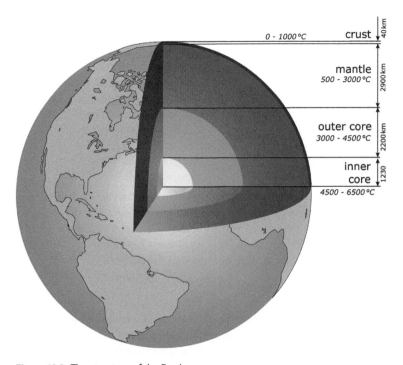

Figure 10.2 The structure of the Earth.

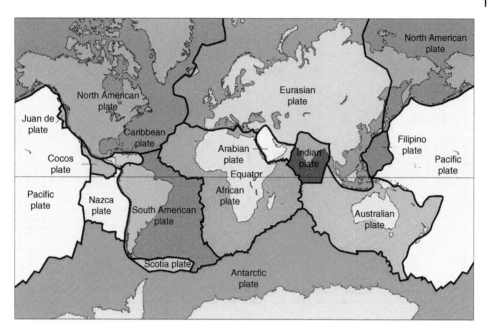

Figure 10.3   Tectonic plates on Earth. Source: US Geological Survey.

occur in areas where two plates collide. Thermal anomalies can also frequently be observed in these areas, where high temperatures can occur, even at shallow depths, creating conditions that enable us to make particularly effective use of the Earth's heat.

Regions like Central Europe, that are not near the borders of tectonic plates do not have optimal geothermal resources. This does not mean that high temperatures do not exist underground. However, compared to regions like Iceland where geothermal conditions are favourable, it takes drilling at greater depths to reach the same temperatures.

The best geothermal conditions in Germany are in the Rhine Rift Valley (Figure 10.4). In this region temperatures of 150 °C or more can be found at depths of 3000 m. The average thermal depth gradient is around 3 °C per 100 m. On this basis, a temperature increase of 90 °C could be expected at depths of 3000 m. In Iceland, temperatures like this occur at depths of just a few hundred metres.

Deep drilling is required to exploit the high temperatures. The technique for this type of drilling has long been familiar from oil extraction. Rotary drilling methods are employed, whereby motors drive a diamond bit into the depths. At very great depths, a drill bit cannot be driven in the conventional way over a drill string due to the increased strain caused by turning and friction. An electric motor or turbine therefore directly drives the drill bit.

The actual derrick (Figure 10.5) that holds the drill pipe is all that can be seen from the surface. Water is pressed into the borehole at pressures of up to 300 bars through the inside of the drill. Crushed rock material in the outer area between the bit and the borehole is forced to the surface by the flushed mud, which at the same time cools the drill. The drilling drive can also be used to change the bore from its vertical position.

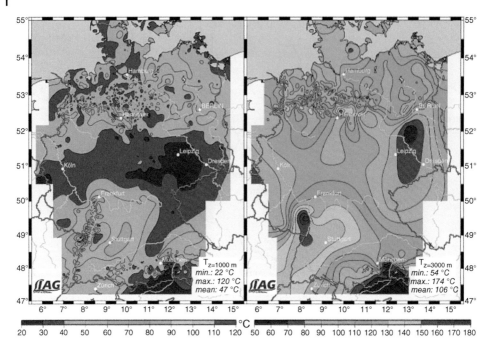

Figure 10.4 Temperatures in Germany at depths of 1000 and 3000 m. Source: http://www.liag-hannover.de [Sch02].

Figure 10.5 Left: New and used drill bits. Right: Structure of a derrick. Photos: Geopower Basel AG.

This enables the boreholes that are sited close together on the surface to exploit a larger area underground.

Depending on the conditions underground, the walls of a borehole can become unstable. Steel tubes are inserted at large intervals and stabilized with a special cement to prevent the walls from collapsing. The drilling is then continued with a smaller drill bit. For many years the very high salt content of thermal water caused serious problems at many drilling sites. Salt attacks metal and very quickly causes corrosion. Today, the use of specially coated materials has eliminated this problem.

The deepest drilling ever carried out was for research purposes on the Kola peninsula in Russia. This bore hole had a depth of 12 km. In Germany, the continental deep geothermal drilling project in the Oberpfalz region reached a depth of 9.1 km. These drilling depths represent the current technical limit. The conditions at depths of 10 km are extreme, with temperatures of more than 300 °C and pressures of 3000 bar. Under these conditions rock becomes plastic and viscous.

Much shallower depths are sufficient for geothermal use. The maximum drilling depth currently planned for large-scale facilities is around 5 km. However, the techniques employed even at these depths are complex and therefore expensive.

A distinction is made between the following in the exploitation of geothermal deposits:

- hot steam deposits
- thermal water deposits
- Hot Dry Rock (HDR)

Hot steam and thermal water sources can be used directly for heating purposes or to generate electricity. If the ground consists only of hot rock, this can heat up cold water that is compressed into the depths.

10.2   Geothermal Heat and Power Plants

10.2.1   Geothermal Heat Plants

If boreholes already exist in a thermal water area, it is comparatively easy to develop a geothermal heat supply. In geothermal heat plants a feed pump fetches hot thermal water from a production well to the surface (Figure 10.6). As thermal water often has a high salt content, along with certain natural radioactive impurities, it cannot be used directly to provide heat. A heat exchanger extracts the heat from the thermal water and transfers it to a district heating grid. A reinjection well injects the cooled thermal water back into the earth.

Relatively low temperatures of 100 °C or less are sufficient for heating purposes. Drilling depths of about 2000 m are often sufficient in suitable regions.

A central heating plant controls the flow rate depending on the heat demand. If the heat demand is particularly high, a peak-load boiler can cover the heat peaks. A backup boiler is also helpful to guarantee reliable heat supply in case problems occur with the extraction pump or the well.

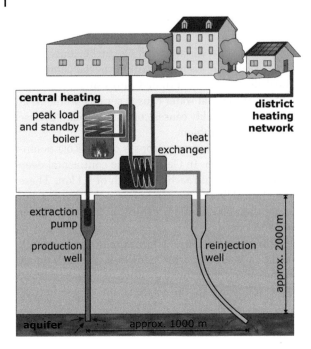

Figure 10.6 Principle of a geothermal heat plant.

10.2.2   Geothermal Power Plants

Using geothermal energy to generate electricity is somewhat more complex than providing thermal heat. The relatively low temperatures that are available call for innovative power plant concepts such as:

- Direct steam use
- Flash power plants
- ORC (Organic Rankine Cycle) power plants
- Kalina power plants

Normal steam turbine systems can be used in geothermally optimal locations with temperatures between 200 and 300 °C. If hot steam deposits exist underground, they can be used directly to drive the turbines.

If hot thermal water is under pressure, it can be evaporated through an expansion stage. The steam from the water that is still hot can in turn be transferred directly to a steam turbine. This technique is called a flash process.

At temperatures of 100 °C or less geothermal water is not hot enough to vaporize water. A normal steam turbine using water as the work medium is not suitable in this case. An ORC system can be used instead (Figure 10.7).

A steam turbine also forms the core of this kind of system. However, instead of water, the steam turbine uses an organic material such as Isopentan or PF5050. A heat exchanger transfers the heat from the geothermal cycle to the organic working fluid. This material also evaporates under high pressure at temperatures lower than 100 °C. The steam from the organic material drives the turbine and expands in the process.

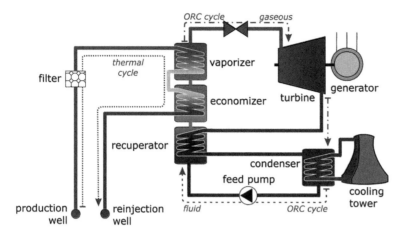

Figure 10.7 Principle of a geothermal ORC plant.

The organic material is liquefied again in a condenser. A cooling tower dissipates the residual heat. A feed pump again puts the organic material under pressure and the heat exchanger completes the cycle.

The disadvantage of ORC systems is that they are relatively inefficient. At temperatures of 100 °C the efficiency is well below 10%. This means that at best 10% of the geothermal heat energy can be converted into electric energy. The Kalina process promises slightly higher efficiency. With this process a mixture of water and ammonia acts as the working material. However, not even the laws of physics can make this process a viable option. Generating electricity from geothermal energy is basically preferable and offers higher efficiency.

The first geothermal power plant in Germany was in Neustadt-Glewe between Hamburg and Berlin (Figure 10.8). This ORC plant, which was built in 2003, had an

Figure 10.8 The Neustadt-Glewe geothermal heat power plant was the first plant to generate electricity from geothermal energy in Germany.

electrical output of 230 kW. A geothermal heat plant, which was put into operation in 1995 with geothermal heat output of 4 MW, is located at the same site. After a technical defect in 2010, power generation was discontinued. Since then, the plant has only supplied heat.

In late 2007 the second geothermal power plant in Germany came into operation in Landau. It uses two boreholes to access 155 °C hot thermal water from a depth of 3300 m. An ORC process designed for continuous all-year-round power generation uses this supply to generate around 3 MW of electricity.

The next power plant was built a short time later in Unterhaching near Munich. Hot thermal water with a temperature of 122 °C is extracted here from a depth of 3350 m. This geothermal cogeneration plant with an electricity output of around 3.36 MW is the first one to use the Kalina process. In addition, the plant feeds heat with a thermal capacity of 38 MW into a district heating grid.

In the meantime, numerous other projects have been developed or are in the planning and construction stages. Information on these projects can be found on the websites provided below.

- www.geotis.de — Geothermal Information System
- www.geothermie.de — German Geothermal Energy Assn.
- http://www.tiefegeothermie.de — Deep Geothermal Energy Information Portal

10.2.3 Geothermal HDR Power Plants

The aim of drilling at depths of up to 5000 m is almost always geothermal power generation. At such depths, temperatures are around 200 °C, even in regions that do not benefit from optimal geothermal conditions, such as Germany, France, and Switzerland. As a result, quite reasonable efficiency is achieved in power generation provided the depth is great enough.

Thermal water cannot usually be exploited at such great depths. What is mainly found is HDR. Artificial shafts are created in which water can be heated to extract the heat from the rocks. These shafts are made by compressing water at high pressure into a borehole. The heat expands the borehole, creates new fissures and expands existing cracks. This produces an underground fracture system that can extend to several cubic kilometres. The activities are monitored via a dedicated auxiliary borehole.

This direct tapping of hot water deposits is also referred to as hydrothermal geothermal energy, whereas hot-dry-rock technology is also called petrothermal geothermal energy.

To generate geothermal energy, a pump transports cold water through an injection well down to depth. There, the water disperses into the cracks and fissures of the crystalline rock and heats up to temperatures of 200 °C. The hot water reaches the surface again through production wells and delivers the heat over a heat exchanger to a power plant process and district heating network (Figure 10.9).

During the 1970s the Los Alamos laboratory in the USA conducted the first tests of the HDR method. A European research project on HDR technology has been underway

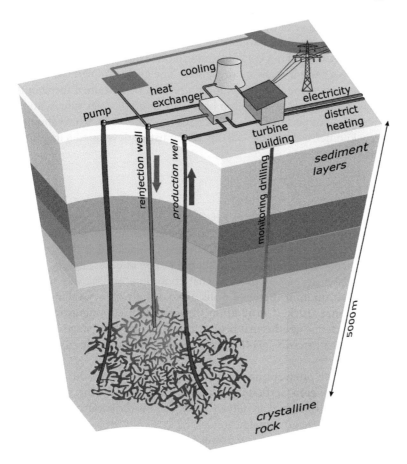

Figure 10.9 Diagram of an HDR power plant.

in Soultz-sous-Fôrets in Alsace since 1987. The power plant started a trial operation in 2008.

In 2004 Geopower Basel AG was founded in Switzerland with the aim of building the first commercial HDR power plant. However, work on the project was stopped in 2007 when small tremors occurred during the creation of the underground fractures.

10.3 Planning and Design

The most important factor in planning geothermal power plants is the temperatures that can be achieved. The design of the heat exchangers, district heating grids and plant processes are all based on projected temperatures. Geologists try to determine in advance at which depths the desired temperatures can be found. To an extent, they are able to rely on the knowledge gained from existing boreholes.

In addition to achievable temperatures, extractable water quantities also play a major role. Large volumes of water are needed to achieve high output. The diameter of the borehole as well as the pumps both have to be designed accordingly. Last but not least, it is important that the temperature of the thermal water is not allowed to drop too significantly during the extraction process. Large-scale power plants usually extract more heat from the depth than regularly flows back into an exploited area. Therefore, a slow cooling of the exploited area cannot really be prevented. The goal is to plan the intervals between the drilling so that the desired temperatures can be sustained for about 30 years. After this period the temperatures drop below the desired target values, and, as a result, the geothermal plant performance also decreases. A new site, which should not be more than a few kilometres from the existing site, must be developed if any further exploitation is planned.

10.4 Economics

Drilling is by far the biggest cost factor with deep geothermal energy. Yet, it is not only the cost of the drilling itself that is the problem. The risks associated with it, especially on commercial projects, are something that should not be underestimated. Even the best geologists can never accurately predict what conditions will be like underground. Drilling costs shoot up immediately if drills unexpectedly hit crystalline hard rock instead of soft sedimentary rock. If, in addition, the temperatures underground are well below what was projected, a geothermal project can fail at an early stage. At Bad Urach, for example, a rather promising geothermal project was repeatedly suspended for financial reasons.

Underground conditions can often spring other surprises. For example, during work on a geothermal project in Speyer in Germany, drillers not only found the thermal water they were hoping for but also discovered an oilfield 2000 m below the surface Now oil is being produced there instead of geothermal heat.

Even when drilling is successful, geothermal power plants have to spend up to half of all investment on the drilling activities. Therefore, the cost of geothermal power is often considerably higher than that of electricity from wind and hydropower systems.

In Germany, the Renewable Energy Sources Act (EEG) also promotes geothermal power generation. In 2018, the statutory payment for geothermal power plants was 25.2 cents kWh^{-1}. The payments for new plants are to be reduced by 5% per year from 2021.

The costs in countries with good geothermal conditions are considerably lower than those in Germany. At drilling depths of a few hundred metres the drilling costs in those areas end up being very low. If, in addition, high temperatures exist close to the surface, some countries such as Iceland can even use geothermal heat to keep their pavements free of ice in winter (Figure 10.10).

10.5 Ecology

Geothermal co-generation plants have little ecological impact. Most of the plant is located underground and is not visible, and therefore does not have a direct negative impact on people or landscape. Only the power plant complex is above ground. Like other heat plants, these also require cooling water for the plant processes. However, water is readily available at most geothermal sites.

Figure 10.10   The Nesjavellir geothermal power plant in Iceland. Photo: Gretar Ívarsson.

What can cause problems are some of the working materials used in the generation of power. For instance, the material PF5050 used in ORC processes has a very high greenhouse potential. If 1 kg of this material reaches the atmosphere, it produces the same greenhouse effect there as 7.5 tons of carbon dioxide. However, alternatives, such as Isopentan, are available.

In the long term, geothermal systems can cause a certain cooling of some localized areas under the ground. However, as far as we are currently aware, this has no effect on the surface.

Relatively little research has been done on the risk of seismic activities. Small tremors with an intensity of up to 3.4 on the Richter scale occurred in 2006 after geothermal drilling for an HDR power plant to a depth of 5000 m in Basel, Switzerland. The tremors caused small cracks to buildings in the region, so work on the project was halted. The geothermal company involved paid compensation for most of the damage.

As long as scientists are unable to predict accurately whether and when tremors can occur during the compression of water, HDR projects will pose a certain risk in densely inhabited regions. Hydrothermal geothermal projects that do not require fissures and cracks to be artificially created are comparatively safe with regards to earthquake risk.

10.6   Geothermal Markets

China, the USA, Iceland, and Turkey are the outright leaders when it comes to geothermal heat use. The USA and the Philippines have the highest power plant output from geothermal electricity generation (Figure 10.11). Providing more than 60% of the country's energy needs, geothermal energy in Iceland constitutes the highest relative share of any country's total primary energy supply. However, since the population of Iceland is only 300 000 and the country is not exactly densely populated, its absolute installed capacity is nevertheless lower than that in other countries.

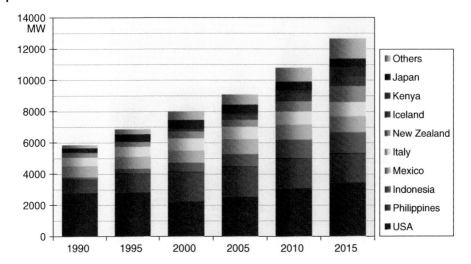

Figure 10.11 Installed geothermal power plant capacity worldwide. Data: IGA, http://www.geothermal-energy.org.

The use of geothermal energy in Germany is considerably more modest than in the countries already mentioned. The main reasons are the comparatively low geothermal resources and the associated great drilling depths for tapping suitable temperatures. However, this disadvantage is also an incentive for technological developments. Germany is one of the leading nations in geothermal power generation in the low-temperature range between 100 and 200 °C.

10.7 Outlook and Development Potential

Many countries are starting to develop the use of geothermal energy, but compared to other renewable energy technologies, such as wind power and photovoltaics, the annual growth rates for this energy source are quite modest.

The share of geothermal energy in the worldwide energy supply is therefore currently very low. But this technology has great potential. Another advantage of geothermal energy is its constant availability. Compared to the fluctuations in certain renewable energy sources – such as solar energy, wind power, and hydropower – geothermal energy is not subject to any unpredictable daily or yearly changes in available supply. This makes geothermal energy an important building block in a carbon-free energy supply. Additionally, supply reliability will become a big factor, as the share that renewable energies have within the overall energy demand increases. This will help to raise interest in building new geothermal plants.

For economic reasons, countries with large geothermal resources will remain the leaders in the field. As prices for fossil fuels continue to rise, geothermal energy will also become more interesting to countries with moderate geothermal resources. Germany could benefit not only from the use of geothermal energy in its own country, but also from the export of technologies developed there.

11

Heat Pumps – From Cold to Hot

Due to the rise in oil and petrol prices during recent years and increasing public awareness of climate and pollution problems, alternatives such as wood pellet heating, solar thermal systems, and heat pumps are becoming more popular. Manufacturers of heat pump systems have recorded strong growth since 2000.

Yet, the whole principle behind the heat pump is far from modern. Lord Kelvin, a British physics professor, already proved this principle in 1852. He also established that a heat pump uses less primary energy to provide heat than a system that produces heat directly. A heat pump uses a heat source with low temperatures and increases it to a higher temperature (Figure 11.1). This process requires an electric, mechanical, or thermal drive.

11.1 Heat Sources for Low-Temperature Heat

A heat pump is basically a machine in which a mechanically or electrically driven pump generates heat from a low-temperature source. This heat is then used to provide space heating or to produce hot water. Before a heat pump can even function, a low-temperature source must be available. The higher the temperature level of the heat source, the more efficiently the heat pump can work.

The following heat sources are available for homes (Figure 11.2):

- Groundwater (water/water);
- Ground, ground heat exchanger/collector (brine/water);
- Ground, ground probe (brine/water); and
- Ambient air (air/water or air/air).

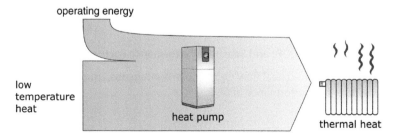

Figure 11.1 Energy flow with a heat pump process.

Renewable Energy and Climate Change, Second Edition. Volker Quaschning.
© 2020 John Wiley & Sons Ltd. Published 2020 by John Wiley & Sons Ltd.

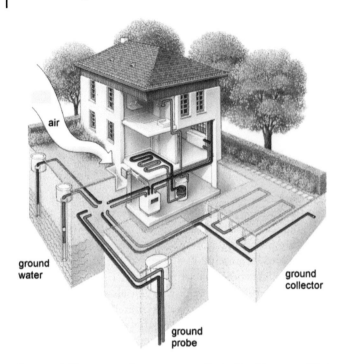

air

ground
water

ground
collector

ground
probe

Figure 11.2 Heat sources for heat pumps. Illustration: Viessmann Werke.

The waste heat from industrial plants can also be used.

Depending on the heat source, heat pumps fall into the categories of air/air, air/water, brine/water, or water/water systems. The heat medium supplied is indicated in front of the forward slash. In the case of ambient air, it is air. In the case of constantly frost-free groundwater, it is water. Because of the risk of frost, a mixture of water and antifreeze, called brine, flows through the pipes in the ground.

The heat medium given after the forward slash is that of the delivered heat. In most cases, heat pumps heat up water for heating and domestic use. They are rarely used to heat the air for air heating systems.

The higher the temperature of the heat source and the lower the temperature needed for heating, the less electric energy is required to drive the heat pump. Due to the low heating temperatures, underfloor heating is preferable to conventional radiators.

Coefficient of Performance (COP) and Seasonal Performance Factor (SPF)

The ratio of instantaneous transmitted heat flow $\dot{Q}_{heating}$ to instantaneous, usually electrical, input P is called the coefficient of performance or COP:

$$\text{COP} = \frac{\dot{Q}_{heating}}{P} = \frac{\dot{Q}_{heating}}{\dot{Q}_{heating} - \dot{Q}_{source}}$$

The electrical input P and the thermal input $\dot{Q}_{source}$ of the low-temperature heat source together produce the heat flow $\dot{Q}_{heating}$:

$$\dot{Q}_{heating} = P + \dot{Q}_{source}$$

For example, if a heat pump with an electrical input of $P = 3\,kW$ generates a heat flow of $\dot{Q}_{heating} = 9\,kW$, the COP is 3. The difference of $\dot{Q}_{source} = 6\,kW$ derives from the low-temperature heat source.

The COP only applies to instantaneous values. The annual average is interesting and is called the seasonal performance factor, or SPF for short.

A high SPF is essential for the environmentally compatible and economical operation of a heat pump. With an SPF of 4, for example, a heat pump can cover a heating demand of $10\,000\,kWh\,yr^{-1}$ using only $2500\,kWh$ of electric energy. With an annual performance coefficient of 2, the electric energy consumption rises to $5000\,kWh$.

Very good systems reach SPF values of around 4. In practice, the values are often below this. Table 11.1 shows typical SPF values for different types of heat pumps from a field test in the Black Forest.

Heat pumps that draw their heat from the ground produced the best values. The SPF values for groundwater heat pumps were somewhat lower. The reason is that it takes more pumping effort to extract groundwater than it does to exploit heat from a closed brine loop in the ground. Furthermore, dirt traps in groundwater wells eventually become blocked, which further increases the amount of pumping energy needed. As ambient air temperatures in winter are lower than ground or groundwater temperatures, air-based heat pumps work least efficiently at that time of the year.

11.2   Operating Principle of Heat Pumps

All heat pumps need a refrigerant contained in a closed loop. The refrigerant absorbs the low-temperature heat. The heat pump then heats up the refrigerant to a higher temperature, which is then utilized. Based on the operating principles, a distinction is made between:

- compression heat pumps;
- absorption heat pumps; and
- adsorption heat pumps.

Table 11.1 Typical annual performance coefficients for electric heat pumps [Lok07]

Heat pump	Heat source	SPF with underfloor heating	SPF with radiators
Brine/water	Ground	3.6	3.2
Water/water	Groundwater	3.4	3.0
Air/water	Air	3.0	2.3

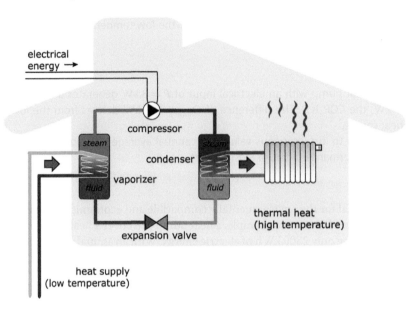

Figure 11.3 Operating principle of a compression heat pump.

11.2.1 Compression Heat Pumps

Compression heat pumps are by far the most common type (Figure 11.3). The principle of these heat pumps is based on a refrigerant with a low boiling point that vaporizes at low temperatures and reaches high temperatures when it is compressed (Table 11.2). The heat supplied from the low-temperature source in the vaporizer is sufficient for vaporization.

Table 11.2 Temperature ranges of common refrigerants

Abbrev.	Name	Boiling point at 1 bar (°C)	Condensation temperature at 26 bar (°C)
R12	Dichlordifluormethane	−30	86
R134a	1,1,1,2-Tetrafluorethane	−26	80
R290	Propane	−42	70
R404A	Mixture of different HFCs	−47	55
R407C	Mixture of different HFCs	−45	58
R410A	Mixture of different HFCs	−51	43
R600a	Butane	−12	114
R717	Ammonia	−33	60
R744	Carbon dioxide	−57	−11
R1270	Propene	−48	61

A compressor (usually electrically driven) brings the vapour-forming refrigerant to a high operating pressure. During this process it heats up considerably. This process is similar to what happens with a bicycle pump when one uses one's thumb to stop the air escaping while energetically pumping a tyre. The heat of the heated refrigerant is then used as useable heat, usually for room heating or heating water. The heat is removed through a condenser that again liquefies the refrigerant. The refrigerant that is compressed expands over an expansion valve, cools off and is transferred to the vaporizer again.

The Reverse Refrigerator

Heat pumps are also used in refrigerators, where they act as refrigerating machines. A vaporizer removes the heat from the interior of a refrigerator. The heat is emitted over cooling fins on the back of the appliance. The heat emitted there comprises the heat removed from the refrigerator and the electrical operating energy of the refrigerator compressor. The refrigerator emits more heat at the back than it extracts from the inside, which is why it is not possible to cool down a room in the summer by leaving the refrigerator door open.

The first compression refrigerating machine was developed by the American Jacob Perkins in 1834. He used ether, which is no longer used today, as a refrigerant for his ice-making machine. The refrigerant ether has the disadvantage that in combination with atmospheric oxygen it forms a highly explosive mixture. This occasionally caused ether ice machines to explode.

11.2.2   Absorption Heat Pumps and Adsorption Heat Pumps

Like compression heat pumps, absorption heat pumps use low-temperature heat to evaporate a refrigerant. However, absorption heat pumps use a thermal compressor instead of the electrically driven compressor used in compression heat pumps (Figure 11.4).

The function of a thermal compressor is to compress and heat the refrigerant. This happens through a chemical process of sorption, for example through the dissolving of ammonia in water. This was explained in the section 'Cooling with the sun' in Chapter 6. The heat released through sorption can be used as thermal heat.

A solvent pump transports the solution to the generator. The solvent pump, unlike the compression heat pump, does not build up high pressure, so a relatively low amount of electrically driven energy is needed. The generator now has to separate the water and refrigerant ammonia in the solution again to enable the sorption to take place once more. High temperature heat is needed for the expulsion. Solar heat and biogas can be used.

The high temperature heat supplied is well below the dissipated quantity of useful heat. The main advantage of absorption heat pumps is that they use a very small amount of valuable electric energy. They are particularly useful for large-scale applications. They are also used as refrigerating machines in refrigerators run with propane gas. The refrigerant ammonia is toxic and flammable. However, it is a widely used chemical and is considered easy to control.

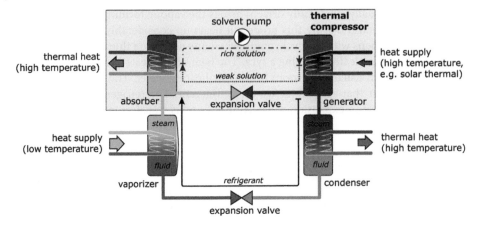

Figure 11.4 Operating principle of absorption heat pumps.

Adsorption heat pumps, which only differ by the second letter from absorption heat pumps, likewise use thermal energy as the operating energy.

What is meant by adsorption is that a gas like water vapour attaches itself to a solid material, such as activated carbon, silica gel, or zeolite. The process of adsorption, i.e. the bonding of the water vapour by the solid material, creates high temperatures that can be utilized by a heat pump. However, adsorption heat pumps are still at the research stage and therefore will not be covered in detail in this book.

11.3 Planning and Design

Today heat pumps are available for almost any thermal output required. Manufacturers usually provide advice on selection and design. The most important aspect in planning is the selection of a low-temperature heat source and how to exploit it.

If a heat pump is to be installed in a groundwater protection area, no groundwater is allowed to be drawn. Approval for the use of a deep-ground probe to utilize the heat from the ground is only given in special cases as the brine could contaminate the groundwater if there were a leak. In Switzerland, for example, there are installations of ground probes in water conservation areas where carbon dioxide (R744) replaces the brine.

Using ambient air as a heat source is the simplest and most cost-effective approach. This is not even a problem in groundwater protection areas. No approvals are necessary to install and operate air/water or air/air heat pumps in these areas. Basically, all these heat pumps need are two openings in a house wall through which the ambient air can be fed to the heat pump. Condensation forms if the outside air is very cold and should be allowed to drain off in a controlled way. Heat pumps can also easily be installed outdoors. Air/water heat pumps function at ambient temperatures as low as −20 °C. A supplementary electric heater helps to ensure that the necessary heat supply is covered when temperatures are particularly extreme. A small buffer storage unit can optimize the operating times of a heat pump. The disadvantage of using ambient air is that the SPFs are relatively low. This significantly increases the electricity consumption compared with other heat sources.

Brine/water heat pumps, in other words heat pumps that extract heat from the ground, use the least amount of electricity. Either ground collectors or ground probes are used to extract the heat. A ground collector usually comprises a set of plastic pipes that are laid in a serpentine pattern in the soil. The optimal depth for the pipes is 1.2– 1.5 m, and the gap between pipes should be about 80 cm.

Size of the Ground Heat Collector

The length l and the area A of a ground collector is calculated from the required refrigerating capacity $\dot{Q}_{source}$ of the low-temperature heat source, the extraction capacity $\dot{q}$ per metre of pipe and the pipe distance d_A:

$$l = \frac{\dot{Q}_{source}}{\dot{q}} \text{ and } A = l \cdot d_A.$$

For example, if the required heat output is $\dot{Q}_{heating}$ 10 kW and the COP is 4, the required cooling capacity $\dot{Q}_{source}$ is 7.5 kW. With dry sandy soil the extraction capacity is around $0.01 \cdot \text{kW m}^{-1}$, with dry clay soil around 0.02 kW m^{-1}. As a result, on the basis of this example, the length of pipe needed for clay soil is calculated as

$$l = \frac{7.5 \text{ kW}}{0.02 \text{ kW/m}} = 375 \text{ m}$$

and with a pipe distance of $d_A = 0.8$ m the collector area is

$$A = 375 \text{ m} \cdot 0.8 \text{ m} = 300 \text{ m}^2.$$

If in doubt, it is recommended to round up the values generously. Since individual pipe lengths should not exceed 100 m, four pipe circuits are recommended, each with a length of 100 m (Figure 11.5).

If sufficient space is not available in a garden, or there is no desire to dig up a whole plot of land, ground probes can be used to tap into the ground heat. Vertical boreholes can reach depths of 100 m. Constant temperatures of around +10 °C or more exist at these depths all year round. U-shaped pipe probes are inserted into the boreholes through which the brine of the heat pump will later flow. The depth of the drilling and the number of probes depend on the heat requirement and the composition of the ground below. Geologists and specialized drilling firms can help with the specifications. Depending on the composition of the ground at the bottom, the potential extraction capacity is between 20 and 100 W m^{-1}. A rough estimate could be calculated at about 55 W m^{-1}. Therefore, about 5.5 kW of refrigerating capacity could be extracted from a 100 m deep probe. In order to achieve a higher output several parallel probes with a distance of at least 5–6 m between them would be necessary.

In addition to the ground heat, heat from groundwater can be used. This requires a production well and a reinjection well. The reinjection well transfers the cooled ground-water back into the ground. It should be sited at least 10–15 m downstream of the production well, in the direction of the groundwater flow, so that the cooled water does not flow back to the supply bore (Figure 11.6).

Figure 11.5 Heat pump installation. Source: Bosch Thermotechnik GmbH.

In many countries, approval must be obtained from the responsible water authority before groundwater can be extracted. This approval is usually given under certain conditions, except in groundwater protection areas. Permission is also often required to drill the ground probes for closed brine/water systems. In Germany, this is issued by the local water authority for drilling depths of up to 100 m. For deeper boreholes, an additional permit from the responsible mining authority is required. Normally, the drilling company applies for the required approvals.

Figure 11.6 Air/water heat pump installed outdoors, without the need for drilling (left). Deep drilling for a water/water heat pump (right). Photos: STIEBEL ELTRON.

From the Idea of a Heat Pump to Owning One's Own System

- Determine possible heat sources:
    Ground probe – Does deep drilling require approval?
    Ground collector – Can a large section of the garden be dug up?
    Groundwater – Is the house located in a groundwater protection area?
    Air – Suboptimal solution if no other sources are available
- Calculate heat demand and heating capacity.
- Can insulation help to reduce the heat demand?
    Can the required temperatures be reduced through underfloor heating or larger radiators?
- Request quotations for heat pumps and, if necessary, for the drilling.
    Tip: Only hydrofluorocarbons (HFC)-free heat pumps should be used for climate protection reasons.
- If necessary, have the drilling company obtain approval for drilling.
- Determine optimal energy tariff, and, if necessary, plan to have a buffer storage unit.
    Tip: Choose a 'green electricity' option for climate protection reasons.
    A photovoltaic (PV) system can cover part of the electricity demand at low cost.
- Examine favourable financing options in the context of wider climate protection measures.
- Arrange for the system to be installed by a qualified company.

11.4   Economics

Investment costs for a typical heat pump installation in a single-family home are between €8000 and €12 000. Added to this are the costs of tapping into a heat source, which are in the order of €3000–6000 for ground collectors or ground probes.

The costs of conventional heating systems are not applicable to new buildings with heat pumps. For example, with gas heating, the costs can include a gas connection and a chimney in addition to the gas boiler. Nevertheless, the investment costs for heat pump systems are usually considerably higher than for conventional gas or oil heating.

In Germany, between 2000 and 2013 heat pumps enjoyed economic advantages but these subsequently stagnated due to sharp increases in mains electricity costs. In particular, the sharp drop in oil prices after 2013 made it more difficult for heat pumps to compete (Figure 11.7). While domestic electricity is heavily burdened with taxes, taxes on heating oil and natural gas are comparatively low, although both energy sources have to be imported and are also questionable from an environmental perspective. It is to be hoped that legislation will intervene and improve the economy of the heat pump at the expense of climate-damaging oil and gas heating.

Due to the enormous cost reductions for PV systems over recent years, the operation of a heat pump with electricity from one's own PV system has been the most cost-effective way of supplying heat since 2012. However, in Germany it is generally not possible to cover the entire heat demand with PV electricity, due to the low solar radiation supply in winter, so in practice there will be a mixed calculation based on expensive grid electricity and low-cost PV electricity. With efficient heat pumps with SPF values of more than 3, in many cases the operating costs are lower than those of oil and gas

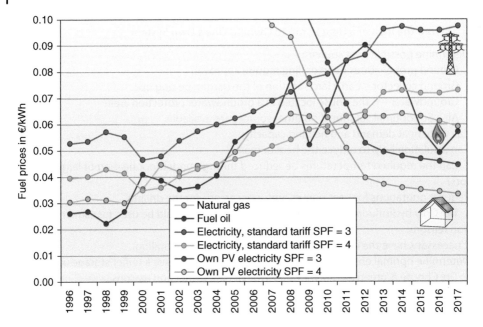

Figure 11.7 Development of domestic prices for gas, oil, and electricity for the operation of heat pumps for different seasonal performance factors (SPF) in Germany.

heating systems. In contrast to the strongly fluctuating oil prices, the combination of a heat pump with a PV system ensures that the heat prices remain stable in the long term.

For large PV systems with a capacity of 10 kW or more, which can be achieved with a PV surface area of 60 m^2 or more, German lawmakers have imposed an additional tax on self-consumption of climate-friendly solar electricity since 2014. This further worsens the economy of heat pump systems power from PV electricity compared to oil and gas heating.

If the heat pump is part of a modernisation measure or contributes to achieving the low-energy house standard, it may be possible to obtain low-interest loans through the German KfW Bank. The Federal Office of Economics and Export Control (BAFA) also promotes the installation of efficient heat pumps.

- www.klima-sucht-schutz.de/ neubau.0.html
- www.kfw.de
- www.bafa.de

Comparison of heating costs in new buildings
KfW Bank
BAFA funding

11.5 Ecology

Heat pumps are generally associated with a positive effect on the environment. But this is not always the case. The Achilles heel of heat pumps is the refrigerant. The range of refrigerants for compression heat pumps is broad. Chlorofluorocarbons (CFCs) were often used during the first heat pump boom. Because of their negative impact on the ozone layer, their use has been banned in new systems.

Today HFC, often called CFC equivalent substitutes, are mostly used. Although they are harmless to the ozone layer, they share another characteristic with CFC that is negative for the environment: both materials have extremely high greenhouse potential. As a result, even small quantities of between 1 and 3 kg of refrigerant in heat pumps for single-family homes are developing into an ecological problem. That is why the EU has restricted the use of climate-damaging refrigerants, with long transitional periods.

If 2 kg of HFC R404A reaches the atmosphere, it develops the same effect on the climate there as 6.5 tons of carbon dioxide. This quantity of carbon dioxide is emitted during the burning of 32 kW h of natural gas. The same amount heats a standard new-build house for three years and a 3-l house for an impressive nine years. The electricity demand of the heat pump itself is not even included in this balance sheet.

If a leakage occurs in a heat pump system, the refrigerants escape quickly because they evaporate under normal environmental conditions. On the positive side, HFC substances are neither toxic nor inflammable. However, in terms of climate-compatibility, the fast volatility of refrigerants turns out to be a problem. Not every heat pump will develop a problem whereby its entire content of refrigerant escapes into the atmosphere. However, refrigerant loss is unavoidable during the filling and disposal of a system and, due to continuous seepage, during regular operation. Nowadays standard heat pump suppliers seldom use refrigerants that can adversely affect the climate. At the same time heat pumps with R290 or propane as the refrigerant are not showing performance data that is any worse that those with refrigerants containing HFC. Special safety measures must be taken due to the flammability of refrigerants R290, R600a, and R1234, but for the most part these are easy to implement. In the past, more pressure to use HFC-free refrigerants was obviously placed on the manufacturers of refrigerators and freezers than on those of heat pumps. As a result, in Europe refrigerants that are not harmful to the environment have become part of the standard range in this area for years, despite their flammability.

In contrast, the HFC problem with heat pumps is still hardly ever discussed. Most manufacturers publicize the HFC they use as being environmentally friendly. One manufacturer is even brazenly claiming on their website that the refrigerant R407C is HFC-free. The layperson will find it almost impossible to distinguish between appliances and devices that contain HFC and those that are HFC-free. Table 11.3 provides some further guidance in this area, while HFCs are still permitted as refrigerants.

Suppliers of heating systems often list heat pumps in the 'renewable energy' category. But this is only correct up to a point. Although most of the useable energy of a heat pump is in the form of renewable low-temperature heat from the environment, the power almost always comes from an electrical socket. This power is delivered by regular energy suppliers that frequently offer special tariff conditions because of the high quantity of electricity purchased for heat pump systems. In many countries this electricity ends up coming from coal-fired or nuclear power plants. In Norway, however, hydropower plants generate almost all the electricity used in the country. This makes the heat pump there a completely renewable system. Some countries offer the option of changing to green energy suppliers. Also, the heat pump can be powered at least partly

Table 11.3 Greenhouse potential of different refrigerants relative to carbon dioxide

Abbrev.	Name	Material group	Greenhouse potential
R12	Dichlordifluormethane	CFC	6640
R134a	1,1,1,2-Tetrafluorethane	HFC	1300
R404A	Mixture of different HFCs	HFC	3260
R407C	Mixture of different HFCs	HFC	1530
R410A	Mixture of different HFCs	HFC	1730
R290	Propane	HFC-free	3
R600a	Butane	HFC-free	3
R744	Carbon dioxide	HFC-free	1
R717	Ammonia	HFC-free	0
R1234yf	2,3,3,3-tetrafluoropropene	HFC-free	4
R1270	Propene	HFC-free	3

by electricity from one's own PV system. In this case, too, a heat pump system is completely renewable and therefore free of direct carbon dioxide emissions.

If conventional electricity rather than green electricity is used to operate a heat pump, the savings in carbon dioxide emissions are significantly lower due to the poor efficiency of fossil thermal power plants compared to modern natural gas heating (Figure 11.8). If the heat pump also uses a HFC refrigerant that impacts the environment, in an extreme case the environmental balance can turn out to be even worse than with a modern heating system using natural gas.

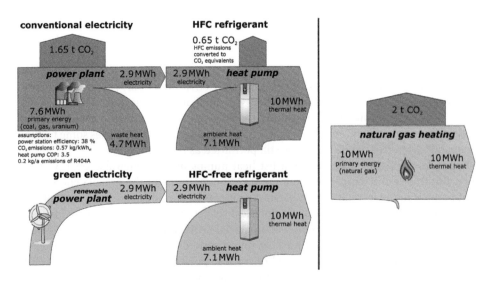

Figure 11.8 Environmental balance of two heat pump heating options and natural gas heating.

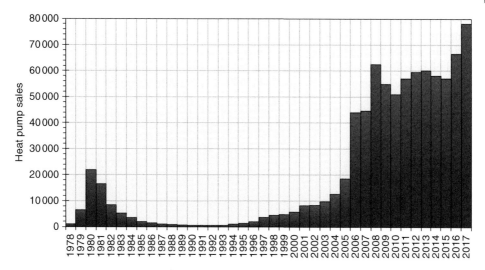

Figure 11.9 Sales of heat pumps in Germany.

11.6 Heat Pump Markets

After the first oil crisis in the 1970s, the heat pump sector experienced a real boom. However, due to technical problems, a drop in oil prices, and a lack of environmental compatibility, the market for heat pumps had collapsed almost completely by the late 1980s.

The market in Germany did not revive until the mid-1990s (Figure 11.9). However, heat pumps are much more popular in certain other countries than they are in Germany. Within the European Union, around one million heat pumps were installed in 2016. The markets in Sweden and Switzerland were significantly ahead of Germany. As the average carbon dioxide emissions resulting from the generation of electricity in Switzerland and Sweden are substantially lower than in Germany, heat pumps in those countries also have a much more favourable environmental balance. Heat pumps are currently used primarily in new residential buildings. Meanwhile, around 30% of all new buildings in Germany are heated with heat pumps.

11.7 Outlook and Development Potential

The environmental balance of heat pumps is improving continuously as a result of the steady increase in the share of renewable energies used to generate electricity. From an ecological perspective, the heat pump will become one of the most important alternatives to conventional heating systems once HFC-free alternatives start to replace refrigerants containing environmentally harmful HFC.

As renewable heating systems, such as solar thermal systems and biomass heating (see Chapter 12), can only cover a certain proportion of the heat demand in many countries, the heat pump is a key component of a carbon-free heat supply system. Buffer storage can also be used to change the operating times of heat pumps. This would enable heat pumps to be partially centrally controlled, thus helping to reduce service peaks in the electricity network. With a high availability of wind power, for example, heat pumps would fill heat storage tanks and then draw the heat again at times when power supplies were low. These possibilities indicate that a further expansion of the heat pump market can be expected.

12

Biomass – Energy from Nature

Humans have been using energy from firewood for at least 790 000 years – ever since Stone Age people discovered how to make fire (Figure 12.1). This makes biomass the oldest renewable energy source by a huge margin. In fact, biomass was the most important energy supply worldwide well into the eighteenth century. Even today some countries like Mozambique and Ethiopia use traditional biomass to cover over 90% of their primary energy needs.

As the use of fossil energy supplies grew, biomass use was almost non-existent in the industrialized nations. In 2000 the share of biomass in the primary energy supply in countries such as Britain, Germany, and the USA was not even 3%.

Biomass started to become popular again even in the industrialized countries when oil prices began rising dramatically at the beginning of the twenty-first century. In addition to its traditional use in the form of firewood, modern forms of biomass are now being exploited. Biomass is not only used in simple open fires, but also to operate modern heating systems and power plants for generating electricity, as well as to produce combustible gases and fuels.

Figure 12.1 People have been using the energy from firewood for thousands of years.

Renewable Energy and Climate Change, Second Edition. Volker Quaschning.
© 2020 John Wiley & Sons Ltd. Published 2020 by John Wiley & Sons Ltd.

Figure 12.2 The sun is responsible for the growth of biomass on Earth.

12.1 Origins and Use of Biomass

The term 'biomass' refers to a mass of organic material. It comprises all forms of life, dead organisms and organic metabolism products. Plants are able to create biomass in the form of carbohydrates through photosynthesis. The energy needed is supplied by the sun (Figure 12.2). Only plants carry out this process; animals can produce biomass only from other biomass. This is why all animals would starve to death without plants.

Origin of Biomass

Through photosynthesis plants convert carbon dioxide (CO_2), water (H_2O) and auxiliary substances like minerals into biomass ($C_kH_mO_n$) and oxygen (O_2):

$$H_2O + CO_2 + \text{auxiliary substances} + \text{energy} \rightarrow \underbrace{C_kH_mO_n}_{\text{biomass}} + H_2O + O_2 + \text{metabolic products.}$$

In the simplest case, so-called oxygenic photosynthesis produces glucose ($C_6H_{12}O_6$):

$$12H_2O + 6CO_2 + \text{solar energy} \longrightarrow C_6H_{12}O_6 + 6H_2O + 6O_2.$$

Almost all the oxygen in the Earth's atmosphere is formed through oxygenic photosynthesis. Therefore, the oxygen we need in order to breathe is a pure by-product of biomass production.

Biomass is distributed in many different forms throughout Earth. In addition to solar energy, water is essential for the growth of biomass. Even the solar energy in the northernmost regions of the world is sufficient to create biomass. Regions with water shortages, however, have low biomass growth (Figure 12.3).

Plants therefore convert sunlight into biomass using natural chemical processes. An efficiency can also be established for this process. As a result, land usage for biomass

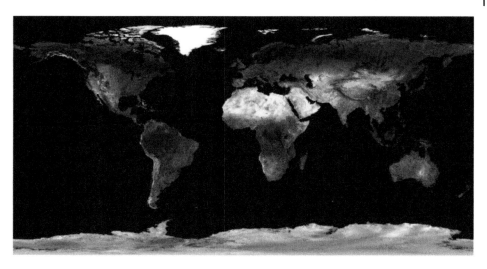

Figure 12.3 The view from space – vast areas of water cover the globe. Source: NASA.

cultivation is comparable to other renewable energy technologies such as solar systems. A plant's efficiency is determined by dividing the calorific value of dried biomass by the solar energy that reached the plant during its growth phase.

On average, including deserts and oceans, the efficiency of biomass production on Earth is 0.14% [Kle93]. Despite the comparatively low efficiency, biomass is created worldwide with an energy content that corresponds to about 10 times our entire primary energy requirements.

Yet not all biomass can be used as energy. Human beings currently use around 4% of new biomass. Two percent goes into food and fodder production and 1% ends up as wood products, paper, or fibre. Around 1% of newly created biomass is used as energy – usually in the form of firewood – and it therefore covers about 10% of the world's primary energy demand.

The plants that reach the highest efficiency during the conversion of sunlight into biomass are C4-plants. These plants have a rapid photosynthesis and, as a result, use solar energy particularly effectively. C4-plants include amaranth, millet, corn, sugarcane, and willow. Under optimal conditions these plants achieve efficiencies of 2–5%.

With biomass a distinction is made between the use of waste from agriculture and forestry and the selective cultivation of energy plants. In Germany studies have shown there is a total potential of around 1200 petajoules per year (Table 12.1). This equated to about 8% of Germany's primary energy needs in 2005. The potential should be similar for other industrialized countries with a high population density. Even if extensive energy-saving measures are implemented, biomass can still only cover part of the energy demand.

There are many different possibilities for biomass use (Figure 12.4). The biggest potential exists with wood and wood products. Waste from agriculture and forestry and biogenic waste are also important from an energy perspective. In addition to waste use, special energy plants can be grown. However, as energy plants compete with food

Table 12.1 Biomass potential in Germany [Kal03]

	Useable quantity in millions of tonnes	Energy potential in PJ/a
Stalk-type (straw, grass)	10–11	140–150
Wood and wood residue	38–40	590–622
Biogas substrate (biomass waste and residue)	20–22	148–180
Sewage and landfill gas	2	22–24
Energy plant mix	22	298
Total biomass potential	92–97	1198–1274

production for arable land, there is some controversy over large-scale cultivation of these plants.

The next step is processing the biomass materials. These materials are dried, compressed, fermented into alcohol, converted into biogas, pelletized or processed into fuel in chemical plants. The aim of the processing is to produce useable biomass fuels.

This biomass fuel has the same spectrum of use as fossil fuels like coal, crude oil, and natural gas. Biomass power plants can use biofuels to generate electricity; biomass heating can satisfy heating needs; and biofuels can be used to run cars and other vehicles.

The versatility of biomass use has led to a real interest in alternative fuels. In many industrialized countries like Germany and Britain, however, biomass falls far short of becoming a complete replacement for fossil fuels. Nevertheless, biomass fuels will play an important role in the renewable energy sector in the future.

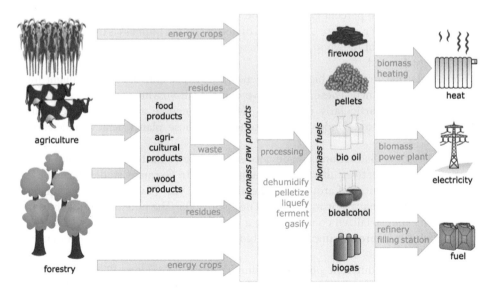

Figure 12.4 Possibilities for biomass use.

12.2 Biomass Heating

With traditional biomass, the focus has been on the generation of heat for cooking and heating. Even today the use of biomass for heating is one of its key applications. Wood, straw, and biogas are the commonly used fuels. Wood, straw, and biogas are the commonly used fuels. Vegetable oils and bioalcohol are also used in some heating systems.

12.2.1 Wood as a Fuel

Wood is by far the main fuel used for biomass heating. It is available in different processed forms (Figure 12.5). As the first step, felled trees are cut to a common length to produce round wood. High-quality woods are not used as fuel but are processed further by the timber industry.

The round wood is then cut up, either by hand or by machine, to produce firewood. Wood scraps or inferior-quality wood may be processed into wood shavings, which can be made into wood briquettes or wood pellets. Special compression techniques are used to press the wood into the right shape for burning. The natural lignin of the wood serves as a binder, so no additional binders are required.

Because of their uniform shape and small size, wood pellets are an ideal fuel. They can easily be delivered in bulk tankers and then blown into special pellet stores. This eliminates the need for time-consuming manual loading. Automated feeding systems enable

Figure 12.5 Different processed forms of wood. From top left to bottom right: round wood, firewood, wood briquettes, wood pellets.

wood pellet heating systems to provide the same level of heat and ease of operation as natural gas or oil heating systems.

Some isolated quality problems existed in the early days of wood pellet production. Pellets that are the wrong size can get stuck in conveyor systems. If the pellets have not been compressed sufficiently, they can disintegrate too quickly and block up a system. Therefore, wood pellets should comply with current standards, such as the EU standard EN 14961-2, which was published in 2010. In addition, there are pellet certification schemes such as the ENplus seal, which not only certifies the quality of the pellet production, but also takes into account retail and logistics aspects. Wood pellets must conform to the following specifications:

- Diameter: 6 ± 1 mm or 8 ± 1 mm, length: 3.15–40 mm
- Calorific value: Hi: 16.5–19 MJ kg^{-1} or 4.6–5.3 kWh kg^{-1}
- Bulk density: greater than 600 kg m^{-3}
- Water content: less than 10%, ash content: less than 0.7% or 1.5%
- The new standard CEN/TS 14961 applies to all future production of wood pellets.

A pile of pellets weighing one ton takes up 1.54 cubic metres of space and has a calorific value of 5000 kW h. This equates to a calorific value of approximately 500 l of heating oil. Therefore, two kilos of wood pellets can replace one litre of heating oil.

ⓘ A Tree – the World Record Holder

An 80-year-old beech tree reaches a height of 25 m (82 ft). The crown of the tree has a diameter of 15 m (49 ft) and contains about 800 000 leaves. If the leaves were spread out lying next to one other on the ground, they would cover an area of about 1600 square metres (0.4 acres). This beech tree supplies the oxygen needs of 10 people and in the process absorbs large quantities of carbon dioxide. The 15 cubic metres of wood of this beech have a dry mass of 12 tons, of which around six tons are bound to pure carbon. This corresponds to the carbon content of 22 tons of carbon dioxide.

Yet even this massive beech is dwarfed by some tree species. The majestic sequoia reaches heights of up to 115 m (377 ft), diameters of 11 m (36 ft) with a circumference of over 30 m (98 ft) and volumes of 1500 cubic metres. Not only do trees reach an impressive size, but they can also live to a very great age. Some pine trees live for 5000 years. There is a special type of pine where the rootstock can even survive for more than 10 000 years. New shoots that sprout from them 'only' live to about 2000 years. As a result, trees are presumed to be the oldest, highest, and heaviest living things on Earth.

If one were to compress wood shavings into a cube with sides each 1 m long, the mass and calorific value of this cube would be considerably higher than a cubic metre of wood pellets. The reason is that a relatively large amount of air is trapped within a pile of pellets.

In technical jargon, the measure of capacity for one cubic metre of solid wood mass without any gaps is also called a solid cubic metre. When round wood or firewood is stacked very neatly, air gaps occur. The measure of capacity in this case is called stacked cubic metre. When firewood is thrown loosely on a pile, the air gaps increase. In terms of measure of capacity, this is referred to as bulk stacked cubic metre (BCM). The various measurements of capacity can be converted approximately into one another, with the exact factors depending on the type of wood and the shape the wood takes:

- 1 solid cubic metre = 1.4 stacked cubic metres = 2.5 bulk stacked cubic metres (BCM)

As we all know, wet wood burns poorly, because the calorific value of wood depends critically on water content. Damp wood is also heavier than dry wood. This means that not only the wood itself but also the water it contains has to be transported. The water evaporates when the wood burns. However, to evaporate, it needs energy – which in turn comes from the wood. The calorific value drops as a consequence.

Either the wood moisture or the water content is used to indicate the drying level. As both quantities have different parameters and their values therefore differ considerably, this can lead to some confusion. The water content describes the proportion of water in wet wood. On the other hand, the wood moisture indicates the mass of water contained in ratio to the mass of totally dry wood. If exactly half the weight of wood consists of water, the water content is 50% but the wood moisture level is 100%.

Completely dry wood with 0% water content is referred to as bone-dry wood. For example, bone-dry beech trees have a calorific value of 5 kWh kg^{-1}. When wood is dried outdoors, its water content reduces to between 12% and 20%. The wood should be chopped as early as possible into small pieces, covered and then left to air dry for at least one year, but ideally two years. Wood that has been allowed to dry in covered areas can even end up with water content of less than 10%. Even with 15% water content, the calorific value of beechwood is still 4.15 kWh kg^{-1}. With freshly cut wood with a water content of 50% the calorific value drops to 2.16 kWh kg^{-1} (Figure 12.6). Consequently, its calorific value is considerably less than half that of bone-dry wood.

This example shows that firewood should be well dried before it is burnt to extract the optimal energy content. The mass-related calorific value per kilogram differs minimally with different types of wood. On the other hand, the volume-related calorific value, thus the calorific value of a solid cubic metre or a stacked cubic metre, varies considerably (Table 12.2). Heavy wood like beech burns longer than light spruce.

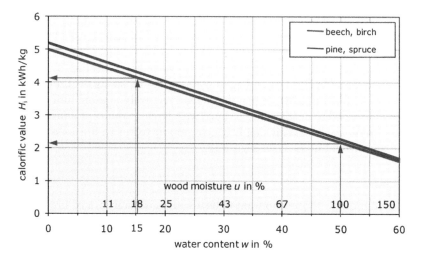

Figure 12.6 Calorific values of wood depending on wood moisture and water content.

Table 12.2 Characteristics of different types of firewood

	Calorific value dried	Density dried	Calorific value Hi with $w = 15\%$		
	$Hi0$ in kWh/kg	In kg/solid cubic metre	In kWh/kg	In kWh/solid cubic metre	In kWh/stacked cubic metre
Beech	5.0	558	4.15	2720	1910
Birch	5.0	526	4.15	2570	1800
Pine	5.2	431	4.32	2190	1530
Spruce	5.2	379	4.32	1930	1350

In addition to providing poor calorific values, too much moisture in wood also has some other undesirable effects. High water content means that wood is being burnt under less than optimal conditions. As a result, it releases a high amount of harmful substances and produces unpleasant quantities of smoke and a pungent smell.

- www.carmen-ev.de   Information on renewable resources
- www.depv.de   Deutscher Energie-Pellet-Verband e.V.

12.2.2   Open Fires and Woodburning Stoves

The classic biomass heating system is the fireplace. For centuries open fires have been used to heat individual rooms. Yet this is a relatively inefficient use of firewood. Open fires usually only reach 20–30% efficiency. This means that 70–80% of the firewood's energy escapes unused through a chimney. As romantic as old castles and palaces may seem, the use of open fires to achieve consistently pleasant room temperatures in draughty castle halls was an almost hopeless task.

With 70–85% efficiency, woodburning stoves are considerably more effective than open fireplaces. A glass panel in front of the stove can be opened to resupply it with wood and then closed after it has been lit. In principle, stoves can also be used to heat up domestic water.

Woodburning stoves are normally only used as supplementary heating systems because of the work involved in cleaning and lighting them, and replacing the firewood.

One of the main problems with stoves is that they need a relatively large amount of air. In addition to the air needed to enable a fire to burn well, large amounts of unused air escape through the chimney. A fresh air supply from outdoors to burn a fire is essential for well-insulated and airtight houses (see Figure 12.7).

12.2.3   Log Boilers

The log boiler is an option for those who want to heat with reasonably priced firewood and without the inconvenience of a stove (Figure 12.8). As these boilers are usually installed in a basement or utility room, they do not have the aesthetic merits of a fireplace. They have large containers for wood supplies that need to be stocked manually, and they can burn for several hours on a single load of wood.

Figure 12.7 Free-standing stove (left) and tiled stove, also referred to as Kachelofen (right). Source: Bosch Thermotechnik GmbH.

Figure 12.8 Solid fuel boiler for heating with logs. Source: Bosch Thermotechnik GmbH.

Compared to the top burn-up of fireplaces where the flame rises upwards, many boilers work with the burn-up at the bottom or on the side. Air is fed into the boiler so that the flame is pointed either downwards or to the side. This increases burning time and cuts down on emissions. The controls, which mainly regulate air supply, ensure that a boiler is burning at an optimal level and adapt it to the heating requirement. Log boilers are available in different performance classes and reach maximum efficiencies of more than 90%. The efficiency of smaller boilers is usually somewhat less.

As log boilers cannot be regulated downward to any specific temperature desired, the installation of a buffer tank is recommended. This will enable a boiler to work under optimal operating conditions at all times. The tank absorbs the excess heat and then continues to supply the heating needed after the firewood has burnt down. The combination of a log boiler with a solar thermal system is a good idea because the boiler can be switched off completely in summer when there is very little need for heating.

12.2.4 Wood Pellet Heating

Wood pellet heating systems offer by far the greatest ease of operation. The fuel is kept in a special pellet store, and an automated feed mechanism using either a feeding screw or a suction device transports the pellets directly to the burner. A screw conveys the pellets from the bottom part of the store. With a suction device similar to a big vacuum cleaner the pellets are also sucked up from below. The suction hoses are very flexible and even enable the bridging of large distances between the store and the burner. As the suctioning of the pellets can be noisy, modern pellet boilers have a small hopper from which the pellets are conveyed to the burner through gravity or a small feeding screw. The hopper is then filled from the store via an automatic timer switch so that the pellet feeding system does not disturb anyone at night.

Ideally, the store would be in a basement. It should be big enough to cover fuel requirements for one year. If a basement is not available, the pellets can also be stored in special silos in a large utility room or in an adjoining shed. Waterproof tanks in the ground are also suitable for stocking pellets. Normally, the store will have two openings (Figure 12.9). The pellets are blown in directly through one opening from the tanker truck that delivers them. The displaced air from the blowing process escapes through the other opening. As the blowing-in of the pellets creates a great deal of dust, a filter traps the wood dust before the air is let out again.

The controls of pellet boilers always ensure that an adequate supply of pellets is available, thereby guaranteeing that the operation of the system is fully automated. The flame is also lit automatically by an electric hot-air blower. When the required heating level has been reached, the heating system switches off again independently. Another feeding screw transports the accumulated ashes to a special ash pan. Wood pellet heating therefore does not require any manual intervention for its daily operation. As with log boilers, buffer storage should be installed to reduce the frequency with which a fire has to be lit.

Wood pellet heating requires a very small investment of the user's time. The soot and ash residue has to be cleaned out of the pellet boiler room every second month or so. In addition, the ash pan should ideally be emptied once or twice a year. The ash can be used as fertilizer in the garden.

In addition to boilers designed for basements or utility rooms, attractively designed pellet boilers, with the flame visible through a glass pane, are available for use in living

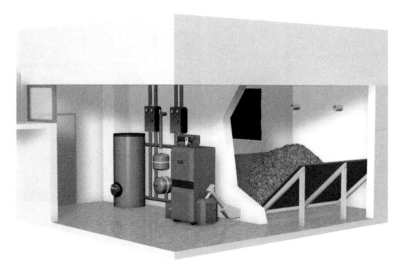

Figure 12.9   Wood pellet heating with pellet store. Source: Bosch Thermotechnik GmbH.

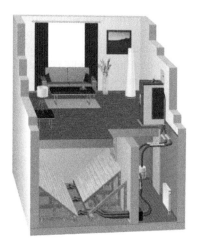 

Figure 12.10   Pellet boiler for living area use with pellet store in a basement. Source: Windhager Zentralheizung.

areas (Figure 12.10). Here, too, a suction device conveys the pellets from the store that would still be located in the basement.

12.3   Biomass Heat and Power Plants

In addition to being burnt in stoves and boilers for heating systems in single homes and apartment blocks, biomass can also be used in large heat plants. A central heat plant consists of a high-performance boiler and a fuel store. The fuel store is usually big enough to ensure that independent operation can be guaranteed for several days or even weeks. A district heating grid then transports the heat to the consumers connected to the grid (Figure 12.11).

Figure 12.11 The Altenmarkt (Austria) biomass cogeneration plant and district heat distributors. Source: Salzburg AG.

With centralized heat plants, individual consumers no longer have to worry about fuel procurement and system maintenance. These tasks are handled by the operator of the heat plant. The efficiency of large heat plants is often somewhat higher than that of small non-central systems. On the other hand, heat losses are higher because of the long pipes in the district heating grid. However, large heat plants fare considerably better when it comes to the emission of harmful substances. Compared to heating systems in single-family homes, large plants use more modern filtering techniques and have stricter conditions on emissions. This ensures that combustion gases emit fewer harmful substances.

Electricity generation is another important application of biomass. Centralized systems that are used exclusively to generate electricity are called power plants. Biomass power plants function in a similar way to coal-fired power plants. The fuels used include wood residue, wood shavings and straw. A steam boiler burns the biomass and produces steam that drives a steam turbine and an electric generator. The principle of steam power plants is explained in the section on parabolic trough power plants in Chapter 7.

Compared to photovoltaic systems and wind turbines, electricity generated by biomass power plants is not dependant on local weather conditions. Biomass fuels are ideal for storage and can be used as and when needed. This makes biomass power plants a viable supplement to other renewable energy plants. They can guarantee electricity supply in situations when there is both little wind or sun available.

In contrast to coal-fired plants that can output more than 1000 mW of power, biomass plants have a considerably lower output in the order of 10–20 mW. The use of biomass fuels is one of the main reasons for this. Biomass fuels usually come from the region where the power plant is located.

Numerous new biomass power plants have recently been built all over the world. One example is the Königs Wusterhausen biomass power plant near Berlin, Germany (Figure 12.12). This power plant has an output of 20 mW and supplies 160 million kilowatt hours per year, which can cover the electricity demand of around 50 000 households. For fuel it uses 120 000 tons of waste and wood residues from the Berlin region annually. The efficiency of this biomass plant is around 35%.

Figure 12.12 The Königs Wusterhausen biomass power plant in Germany (left) and a wheel loader for biomass fuel transport (right). Source: MVV press photo.

Apart from its use in power plants, biomass is also appropriate for cogeneration or combined heat and power (CHP) plants. In addition to electricity, these plants produce heat which district heating grids distribute to consumers. CHP plants generate both power and heat, which is known as cogeneration. CHP plants tend to utilize the biomass fuel more effectively than pure power plants that only generate electricity. The important thing is that a buyer can always be found for the heat that is generated. In summer the heat is often not needed, and in this case, a CHP plant may operate under less favourable conditions than a plant that is optimized for power generation.

CHP plants can be smaller, so that they can be suitable for industrial buildings and even individual houses or apartment blocks. They are usually offered as cogeneration units with a modular configuration. However, the efficiency of smaller systems is frequently lower than that of larger, centralized systems. In addition to solid fuels like wood shavings and pellets, cogeneration units use biofuels and biogas.

12.4   Biofuels

Liquid and gaseous biofuels are more versatile than wood. In addition to generating heat and electricity, biofuels can be used directly as fuel in the transport sector, replacing petrol and diesel. Production methods are available that can convert different biomass raw products into biofuels. Unlike with food production, the prefix 'bio' in this case does not stand for controlled organic cultivation with minimum effect on the environment. On the contrary: the raw materials for biofuels are usually produced using conventional farming methods.

12.4.1   Bio-oil

The biofuel that is the easiest to produce is bio-oil. Over 1000 oleaginous plants can be used for the production of bio-oil. The most popular ones are rapeseed oil, soya oil,

Figure 12.13 Oil-rich plants such as rapeseed and sunflowers are raw material for vegetable oil. Source: Left: Günter Kortmann, North Rhine-Westphalia Chamber of Agriculture.

and palm oil (Figure 12.13). Oil mills produce the vegetable oil directly through either pressing or extraction processes. The residue from the pressing can be reused as animal feed.

Very few older pre-combustion chamber diesel engines can be run on vegetable oil unless they have been converted. Even engines like the Elsbett engine, which were specifically developed to run on vegetable oil, have not yet achieved any significant market share. Vegetable oil is somewhat tougher than diesel fuel and needs higher temperatures to ignite. However, even normal diesel engines can be adapted and converted to run on vegetable oil.

12.4.2 Biodiesel

Biodiesel comes closer to having the characteristics of conventional diesel fuel than pure vegetable oil. Vegetable oil and animal fat are the raw materials used to produce it. The Belgian G. Chavanne applied for a patent for a method to produce biodiesel as early as 1937. Chemically, biodiesel is fatty acid methyl ester (FAME).

In Central Europe rape is normally used to produce biodiesel. Oil mills extract the raw rapeseed oil from rapeseed. Rapeseed meal is a by-product, which usually ends up in the animal feed industry. Rapeseed oil methyl ester (RME) is then created from the rapeseed oil in a transesterification facility.

Production of Rapeseed Methyl Ester (RME)

For the production of RME, rapeseed and methanol together with a catalyst such as a caustic soda solution are placed in a reaction vessel at temperatures of about 50–60 °C. This produces the desired RME as well as glycerine:

$$\text{rapeseed oil} + \text{methanol} \xrightarrow{\text{catalyst}} \text{glycerin} + \text{RME(biodiesel)}$$

Figure 12.14 Corn and other cereals are raw materials for bioethanol production. Source: Günter Kortmann, North Rhine-Westphalia Chamber of Agriculture.

Biodiesel can be used as a substitute for fossil diesel fuels based on crude oil. It is available at many petrol stations in Germany. The engine manufacturer's specifications should stipulate whether a vehicle can run on pure biodiesel. If an engine is not designed for use with this fuel, there is a danger that the biodiesel will eventually destroy the hoses and seals and cause engine damage. Small quantities of biodiesel can be mixed with conventional diesel without a problem, even without approval by the manufacturer. In 2016, biodiesel accounted for 3.2% of the final energy consumption in the transport sector. However, the positive environmental impact of biodiesel is not undisputed.

- www.ufop.de, http://www.ebb-eu.org   Union for the Promotion of Oil and Protein Crops
- http://www.bio-kraftstoffe.info, www.ieabioenergy.com   Information from the Agency for Renewable Resources

12.4.3 Bioethanol

Sugar, or glucose, starch and cellulose are used to produce bioethanol. The raw materials include sugar beet, sugar cane, and grains (Figure 12.14). Sugar can be fermented directly into alcohol. On the other hand, starch and cellulose must first be broken down.

Production of Ethanol from Glucose

Glucose can be converted directly into ethanol in a hermetically sealed environment through fermentation with yeast:

$$C_6H_{12}O_6(\text{glucose}) \xrightarrow{\text{fermentation}} 2CH_3CH_2OH(\text{ethanol}) + 2CO_2(\text{carbon dioxide})$$

Carbon dioxide is a by-product of this reaction. As plants absorb carbon dioxide again during their growth, this reaction does not actually emit any greenhouse gases. The result of the fermentation is mash with an ethanol content of around 12%. Raw alcohol with a concentration of over 90% is extracted through a distilling process. Dehydration over molecular sieves then finally produces ethanol with a high degree of purity. The waste from the ethanol production can be processed into animal feed. The amount of energy required to extract the alcohol is relatively high. If this energy comes from fossil fuels, then the effects of bioethanol on the climate are not favourable. In extreme cases, the impact can even be negative.

Bioethanol can easily be mixed with petrol. An E number indicates the ratio of the mixture. E85 means that the fuel content is 85% bioethanol and 15% petrol. In some countries small quantities of bioethanol are added to petrol. This is not a problem if the ethanol portion is 5% or less. Normal petrol engines can even be run with an ethanol portion of 10% (E10) without requiring any modification, although they should be approved by the manufacturer for this purpose. Engines must be modified for the use of ethanol for anything above that level.

In Brazil flexible-fuel vehicles are very popular. These automobiles can be filled with different mixtures including an ethanol portion of between 0% and 85%. Production facilities that use rye, corn, and sugar beet as raw materials in bioethanol production have also been established in other countries. However, the recent sharp rise in food prices has significantly worsened the economic efficiency of bioethanol production.

12.4.4   BtL Fuels

In the case of pure vegetable oil, biodiesel, and bioethanol, only the parts of a plant that are rich in oil, sugar, or starch can be used to extract fuel. The second generation of biofuels is aimed at overcoming this drawback. The abbreviation BtL stands for 'biomass-to-liquid' and describes the synthetic production of biofuels. Various raw materials, such as straw, biowaste, wood residue, and special energy-rich plants, can be used in their entirety in this process. The result is a major increase in the potential and possible land area yield for the production of biofuels.

The production of BtL fuels is relatively complex. The first stage is gasification of the biomass raw materials. Through the addition of oxygen and steam a synthesis gas consisting of carbon monoxide (CO) and hydrogen (H_2) is produced at high temperatures. Various gas purification stages separate out carbon dioxide (CO_2), dust, and other impurities such as sulphur and nitrogen compounds. A synthesis process converts the synthesis gas into fluid hydrocarbons.

The best-known synthesis process is the Fischer-Tropsch process developed in 1925. Named after its developers Franz Fischer and Hans Tropsch, this process is carried out at a pressure of around 30 bars and temperatures above 200 °C using catalysers. During the Second World War this process was widely used in oil-poor Germany to extract much sought-after liquid fuels from coal. A different procedure then produces methanol from the synthesis gas and processes it further into fuel. During the final production processing stage, the liquid hydrocarbons are separated into different fuel products and refined (Figure 12.15).

BtL fuels have not yet reached a stage where they are ready for mass production. Various companies are currently experimenting with prototype facilities for producing synthetic biofuels. The main advantage of BtL fuels is that they can replace conventional fuels directly, without the need for any engine modifications. However, BtL

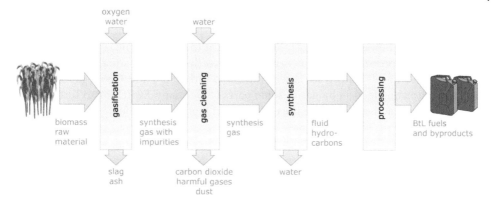

oxygen water	water			
gasification	gas cleaning	synthesis	processing	
biomass raw material	synthesis gas with impurities	synthesis gas	fluid hydro-carbons	BtL fuels and byproducts
slag ash	carbon dioxide harmful gases dust	water		

Figure 12.15 Principle of the production of BtL fuels.

fuels are comparatively expensive because of the complex production procedures involved.

12.4.5 Biogas

In addition to its use in the production of liquid fuels, biomass can also be used in biogas plants to produce biogas. In this process bacteria ferment biomass raw materials in a moist, hermetically sealed environment. The centrepiece of a biogas plant is the heated fermenter (Figure 12.16). A stirring device mixes the substrate and ensures that homogeneous conditions exist. The biological decomposition process mainly converts the biomass into water, carbon dioxide, and methane. The biogas plant captures the gaseous components. The biogas extracted in this way consists of 50–75% combustible methane and 25–45% carbon dioxide. Other components include steam, oxygen, nitrogen, ammonia, hydrogen, and sulphur hydrogen.

In subsequent stages the biogas is purified and desulphurized. It is then stored in a gas storage tank. The biogas yield varies considerably depending on the type of biomass substrate. Whereas with cow dung the gas yield is around 45 cubic metres per ton, with corn silage the yield is at least 200 cubic metres per ton.

Figure 12.16 Biogas plant in a cornfield and view of interior with stirrer. Source: Schmack Biogas AG.

Biogas is used mostly in internal combustion engines. Petrol engines and modified diesel engines are both appropriate. If the engine drives an electric generator, it can produce electric power from biogas. The waste heat from the engine is also useable.

After further processing, biogas can be fed directly into the natural gas grid. This requires the removal of any trace gases, water, and carbon dioxide. Green gas suppliers are now marketing feed-in biogas in some parts of Germany. Switching to this type of gas supplier will actively promote the expansion of biogas production.

12.5 Planning and Design

The possibilities for using biomass are extremely diverse, and would merit an entire book all to themselves. Therefore, this section confines itself to the planning and design of log boilers and pellet heating systems. Both systems are especially relevant to single-family homes.

12.5.1 Log Boilers

Certain facts need to be determined before a log boiler is installed. Wood-fired systems may save on carbon dioxide, but they release harmful substances such as dust and carbon monoxide. As a result, the use of solid fuels is regulated in some cities and towns. Your local chimney cleaning company can be contacted to clarify these conditions. A log boiler also requires sufficient storage area for firewood and a suitable chimney for operation.

If no impediments to the installation exist, the next step is deciding on the dimensions of the log boiler system. The performance of the boiler should at least comply with the rated building heating load, in other words the maximum expected heat demand. A boiler has to run constantly when outdoor temperatures are very low. It is best to have a boiler large enough so that it only has to be filled up once a day under average heating conditions.

Boiler Output and Buffer Storage Size for Log Boilers

The minimum rated boiler output $\dot{Q}_O$ results from the rated burning period of one re-stocking of fuel T_B and the rated building heating load $\dot{Q}_H$:

$$\dot{Q}_O = \dot{Q}_H \cdot \frac{6.4}{T_B}.$$

With a building heating load of 10 kW, which approximately corresponds to that of an average single-family house that has been upgraded with energy efficiency in mind, the rated output of a boiler based on a rated burning period of 2.5 hours is:

$$\dot{Q}_O = 10 \quad kW \cdot \frac{6.4 \ h}{2.5 \ h} = 25.6 \ kW.$$

The required buffer tank volume V_T can be approximated from the rated boiler output $\dot{Q}_O$ and the rated burning time T_B:

$$V_T = \dot{Q}_O \cdot T_B \cdot 13.5 \ \frac{1}{kWh}.$$

As a result, the buffer tank can absorb the total heat quantity of a boiler with a fully stocked combustion chamber. A rated boiler output of 29 kW and a rated burning time of 2.5 hours results in a tank volume of

$$V_T = 29\,kW \cdot 2.5\,h \cdot 13.5\frac{1}{kWh} = 979\,1.$$

12.5.2   Wood Pellet Heating

The general guidelines on log-based heating systems can also be applied to wood pellet heating. As pellet heating systems normally transfer fuel automatically to the burner, there is no loss of comfort due to the frequent firing-up of the system during the day. The rated boiler output can therefore be designed specifically for the heating energy demand of a building. A buffer tank is still useful as it prevents a boiler from firing up too frequently and ensures that the heating system is working at its rated load most of the time. The system then burns at an optimal level and produces a minimal emission of noxious substances.

Unlike split logs, wood pellets are usually stored in a special area within a building. Pellets should not be stored outdoors, as they will absorb moisture and may become damaged. An area of three to five square metres should normally be sufficient as a pellet store for a standard single-family new-build home. Older buildings need more space, whereas well-insulated houses may require considerably less.

The transport system for the pellets is on the floor of the store (Figure 12.17). A slanted shelf will ensure that the pellets slide down to the transport system even when the filling level is low. However, slanted shelves create a void that reduces the useable storage volume to about two-thirds of the space volume. As the price of wood pellets varies throughout the year, the store should hold at least a year's worth of fuel. Pellets can then always be bought at times when prices are low.

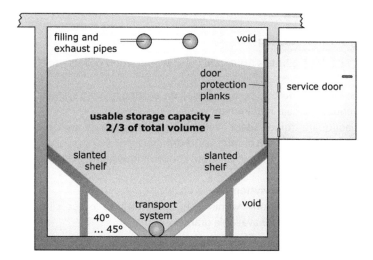

Figure 12.17 Cross-section of a wood pellet store.

Size of the Wood Pellet Store

If the annual heat demand Q_H and the boiler efficiency η_{boiler} are known, the storage room volume V_{store} including any empty space can be calculated:

$$V_{store} = \frac{Q_H}{\frac{2}{3} \cdot \eta_{boiler} \cdot 650\,\frac{kg}{m^3} \cdot 5\,\frac{kWh}{kg}}.$$

If the annual heat demand is not known, an estimate can be made on the basis of 200 kWh per square metre of living area for an average building, 70 kWh m^{-2} for a standard new-build according to EnEV 2009 (50 kWh m^{-2} according to EnEV 2013) and 30 kWh m^{-2} for a 'three-litre' low-energy house, based on Central European climatic conditions. In addition, there is the heat demand for heating water.

According to EnEV 2009, for a new building with 130 m^2 living space, for example, this results in an annual heating requirement of 9100 kWh. Assuming 2000 kWh for hot water, the total heat demand is 11 000 kWh. With an average boiler efficiency η_{boiler} of 80% = 0.8, this results in a tank volume of

$$V_{store} = \frac{11100\ kWh}{\frac{2}{3} \cdot 0.8 \cdot 600\,\frac{kg}{m^3} \cdot 5\,\frac{kWh}{kg}} = 6.9\ m^3.$$

A storage space with a floor area of 2 m × 2 m and a height of 1.73 m would be sufficient in this case. Delivery charges per ton are generally lower for larger quantities. It is important that optimal storage conditions exist so that the pellets do not suffer long-term damage due to high air humidity. ◾

- www.pelletheat.org   Further information available from
  Pellet Fuels Institute

Steps to Installing Bio Mass Heating

- Determine type of fuel. Split logs, perhaps using one's own wood. Wood pellets – for fully automated operation.
- Determine heat demand and heat output, perhaps based on existing heating system.
- Can insulation help to reduce the heat requirement?
- Dimension of store – is adequate storage space available? Is sufficient space available for a buffer tank? Is the chimney suitable for a biomass system?
- Perhaps consult a chimney sweep regarding regulations concerning system operation and residue removal.
- Request quotations for biomass heating systems.
- Examine favourable financing conditions within the framework of other climate-protection measures, check and apply for available grants.
- Arrange for system to be installed by a qualified company.

12.6 Economics

Trying to predict the long-term economic development of biomass fuels compared to fossil fuels is a bit like reading tea leaves. This is illustrated by the fluctuation in wood pellet prices over recent years (Figure 12.18). Whereas in 2003 the prices for heating oil and wood pellets of a comparable calorific value were practically the same, in 2005 oil prices soared by 50%. This caused a boom in the demand for wood pellets, which the industry had a hard time in achieving. In late 2006 the prices for wood pellets were even higher than comparable crude oil prices for a short time. The prices normalized a few months later, and oil prices again rose sharply. During the economic crisis at the beginning of 2009 and in 2015 a similar thing happened.

The potential for producing wood pellets is far from sufficient to supply the entire current heating market. If more and more customers start using wood pellets as fuel, the result will inevitably be a rise in prices. However, as fuel oil prices will also continue to move upwards in the long term, the price advantage of wood pellets could be maintained at a rising level.

Whether wood pellet heating makes economic sense depends primarily on the price difference compared to fuel oil and natural gas. An estimate for installing a wood pellet heating system is around €15 000, which is considerably more than it would be for oil or natural gas heating. However, there are no costs for the natural gas connection in a new building. The cost advantage of wood pellets compared to oil and gas is in the lower running costs of pellet heating. Depending on usage and the price development of fuels, wood pellet heating can be paid back in just a few years. The fuel prices for log-based heating systems are lower than for pellet systems. The operating costs are therefore even lower.

The German Federal Office of Economics and Export Control (BAFA) promotes the installation of automatically fed biomass boilers and log wood gasification boilers in

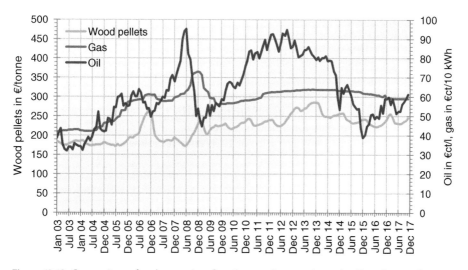

Figure 12.18 Comparison of end user prices for oil, natural gas, and wood pellets. Source: German Federal Statistical Office and German Energy Pellets Association.

Germany. However, the current funding conditions vary relatively frequently and can be found on the internet. If biomass heating is part of a refurbishment or an ecological new building project, in certain cases a low-interest KfW loan can also be taken out for financing.

• www.bafa.de	German Federal Office for Economic Affairs and Export Control
• www.kfw.de	KfW development bank
• http://www.carmen-ev.de	Current prices for biomass fuels

The price development of biofuels in Germany is clearly heading for increases. Until 2006, biofuels such as vegetable oils or biodiesel were exempt from the mineral oil tax. As a result, their price at the pump was significantly lower than that of conventional diesel or petrol. However, with the spread of biofuels, tax shortfalls also increased. Tax rates for biofuels and fossil fuels were therefore adjusted until 2012. By then, the tax rate for vegetable oil and biodiesel had risen to 45 cents per litre. A cost advantage of these fuels is therefore no longer given. In order to avoid bringing the market for biofuels to a complete standstill, a blending obligation for biofuels to gasoline and diesel has been in force in Germany since 2007.

The demand for biofuels is clearly increasing worldwide due to high oil prices and the simultaneous increase in energy demand. This is also putting additional pressure on food prices. Grain and corn prices have reached new record levels in recent years. This development has also raised some ethical issues. Is it right that the quantity of food products being processed into biofuels is increasing when more and more people are unable to afford basic food as it is? An alternative is offered by second-generation biofuels such as BtL fuels and biogas, which can be produced from the non-edible parts of plants.

In Germany, the Renewable Energy Sources Act regulates the remuneration rates for biomass power plants. The tariffs can differ considerably depending on the biomass fuel used and the power plant output. In 2012, the feed-in-tariffs ranged from 3.98 cents per kilowatt hour for a 10-mW power plant using methane up to 14.3 cents per kilowatt hour for a 150-kW biomass power plant. Small anaerobic sewage digestion plants were even awarded 25 cents per kilowatt hour. The tariff is based on the year a plant starts operation and is then valid for 20 years. The remuneration rates for newly constructed landfill, sewage, and mine gas power plants are reduced by 1.5% per year and for biomass power plants by 2.0%. Tenders for biomass power plants started in 2017 and the first tender in 2017 resulted in an average price of 14.3 cents per kilowatt hour.

12.7 Ecology

Biomass has also come under fire for ecological reasons. For example, a farmer in Indonesia who sets fire to a hectare of rainforest in order to grow palm oil for biodiesel to sell to Europe or North America is certainly not helping to protect the climate. Priority should only be given to the sustainable production of biomass raw materials if biomass really is to offer a long-term ecological alternative to fossil fuels.

12.7.1   Solid Fuels

As explained earlier, biomass use is carbon-neutral. During its growth biomass absorbs as much carbon dioxide as it releases again when it is burnt. However, the prerequisite is that the use of biomass is sustainable. Therefore, the amount of biomass that is used should be no higher than what grows back again.

In many countries, solid biomass fuels such as wood and straw usually come from forestry or grain farming in nearby areas. Although the felling of trees, their transport, and finally the processing into fuel creates indirect carbon dioxide emissions, these emissions are comparatively low. For cut wood from the direct vicinity they are almost zero. If one takes into account the indirect carbon dioxide emissions that result from the production and the transport of wood pellets in an overall balance sheet, the carbon dioxide emissions from wood pellet heating are still around 70% lower than those of natural gas and more than 80% lower than those of oil heating (Figure 12.19).

The harmful substances that build up when biomass is burnt create a far greater problem than indirect carbon dioxide emissions. Whereas large biomass power plants have sophisticated filtering systems, single-family homes mostly use their heating systems without any filtering mechanism. The blackened chimneys of their fireplaces testify to this. Even if the carbon dioxide balance sheet turns out to be on the plus side, biomass heating can release all sorts of other harmful substances into the environment if it is not burnt properly. In Germany today, the emission of harmful fine dust from wood-burning plants is already of the same order as that of motorized street traffic. Yet there are clear differences depending on the type of heating system used. Due to their poor efficiency, open fireplaces generally cause particularly high emissions. Therefore, the use of open fires is banned in many places.

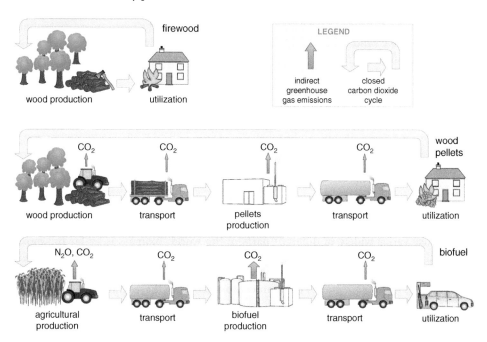

Figure 12.19 Environmental balance sheet for the use of biomass fuels.

When dry firewood or standardized wood pellets are used in modern boilers, the emission values with the same heat output can be 90% lower than with a fireplace. Different companies are now working on producing filters to prevent the emission of dust from small systems.

12.7.2 Biofuels

The ecological impact of biomass fuels is even more controversial. Tractors and farm machinery emit carbon dioxide just by being turned on. Added to this is the amount of energy required to produce fertilizer and pesticides. Nitrogen fertilizers increase nitrous emissions, which are harmful to the environment. Even the processing of biomass raw materials to produce biofuels is energy-intensive. Huge quantities of carbon dioxide are created if the energy needed comes from fossil energy sources.

The real difficulty in evaluating biomass fuels and their ecological impact is illustrated by the example of biomass cultivation in tropical rainforest regions – a topic that has recently been a target of criticism. When tropical rainforests are cleared to cultivate biomass, considerable quantities of carbon dioxide are released through the usual slash and burn methods. The subsequent cultivation of raw materials for the extraction of biofuels then has a negative environmental impact for many years. In other words, from an environmental perspective, it would have been better to burn crude oil from the start.

On the other hand, the results of the production of bioethanol on existing farmland in Brazil are better than in Germany or the USA. Factories in Brazil mostly burn the sugarless residue of sugar cane and extract the energy from this for ethanol production. As a result, ethanol production there is largely carbon-neutral. Other countries use substantial quantities of fossil fuels to do the same thing. This can totally cancel out the climate benefits of bioethanol.

Another critical point with biomass fuels is their limited development potential. If all the available land in the world were devoted to growing biomass for biofuels, it would still probably not be enough to allow biofuels to replace total oil requirements. The second-generation fuels would not totally invalidate this argument but would at least defuse it. If one removes one's own energy needs for fuel production, the net yield per hectare with BtL fuels is around three times higher than with biodiesel (Figure 12.20). The land utilization of solar systems is clearly more efficient than this. In Germany and Great Britain a photovoltaic system with 15% efficiency can generate around 495 000 kW h of electric power per year on 1 ha of land or 200 000 kilowatt hours per acre. This corresponds to a converted diesel equivalent of more than $50\,000\,l\,ha^{-1}$.

12.8 Biomass Markets

Biomass use varies widely in different parts of the world. Biomass is by far the most important energy source in the poorest countries of the world and in some countries even constitutes more than 90% of primary energy needs (Figure 12.21).

The reasons for this are mainly economic. Most people in these countries simply cannot afford crude oil, natural gas, or electricity from coal-fired power plants. In most industrialized countries like Britain, Germany, and the USA the biomass share is well below 10%. The only exceptions are countries like Finland and Sweden, which are densely wooded and sparsely populated.

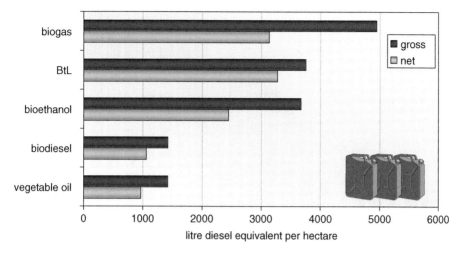

Figure 12.20 Fuel yield per hectare for different biofuels (1 hectare = 2.47 acres). Source: [FNR06].

Biomass is used very differently by industrialized and developing countries. In many developing and emerging countries biomass is still used in the traditional way.

These countries see the main use of biomass as firewood for cooking. Industrialized countries, on the other hand, focus on so-called modern forms of biomass use. These include convenient biomass heating systems, biomass power plants, or biofuels. The proportion of modern biomass is also increasing in industrialized countries. For example, Sweden had set itself a target of becoming independent of crude oil by 2020 mainly through an increase in biomass use but it is not on course to meet this target.

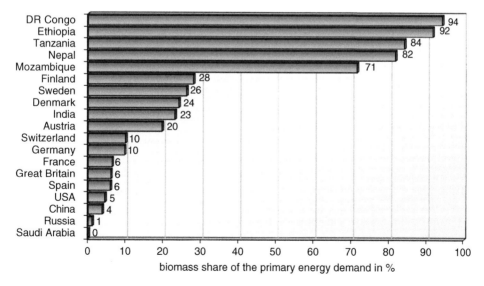

Figure 12.21 Biomass share of primary energy demand in different countries, incl. waste, 2015. Source: [IEA18].

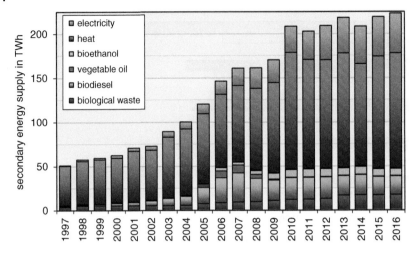

Figure 12.22 Development of biomass use in final energy supply in Germany. Source: [BMWi17].

The potential is lower in more densely populated countries like Germany and Britain. But even in these countries a 10% biomass share in primary energy requirements is conceivable in the long term. Between 1997 and 2010 the biomass share of end energy demand in Germany approximately quadrupled, and now makes up almost 10% of the total energy demand (Figure 12.22).

12.9 Outlook and Development Potential

As the economies of the developing countries grow, the contribution of traditional biomass sources to their energy supply will fall. On the other hand, high oil prices and pressure to act on environmental protection measures will drive the use of modern biomass. Biomass is capable of directly replacing fossil energy sources without needing other technologies to do so. However, the potential of biofuels is not sufficient to replace oil completely. Furthermore, the eco-balance of some biofuels is hotly disputed. However, the potential of biofuels is not sufficient to replace oil completely. There are also ethical concerns about the growing competition with food production. BtL fuels and biogas could be an alternative on a smaller scale.

Biomass power plants will become increasingly important in meeting electricity demand, not least because these plants are able to compensate for some of the major fluctuations in output provided by wind turbine and photovoltaic systems. Biomass will also play a bigger role in providing heat energy. It will therefore remain one of the most interesting renewable energy sources for the future. Throughout the industrialized countries biomass has the potential to achieve a double-digit share of energy supply. Further developments in modern filtering systems are needed to address the problem of harmful substances and fine dust in biomass burning. However, special attention should be given to the sustainable use of biomass raw materials. This is the only way in which it can make a noticeable contribution towards protecting the climate.

13

Renewable Gas and Fuel Cells

Generations of schoolchildren have been entertained by the oxyhydrogen reactions in chemistry class. When hydrogen is oxidized by oxygen from the air, the hydrogen gas explosively releases its stored energy. A spark is enough to ignite a mixture of hydrogen and normal air. In contrast to many other combustion processes, however, the reaction product is absolutely harmless from an environmental point of view. Hydrogen and oxygen simply react to produce pure water.

The proportion of electricity used in energy supply systems is increasing all the time. Fuel cells or gas-fired power plants can generate the much sought-after electricity from hydrogen. The only waste created is water. It is no wonder that the many people with a vision of a global hydrogen economy see it as the solution to our current climate problems. Hydrogen as a single energy source could at the same time help us to get rid of air pollution, acid rain, and other environmental problems caused by the use of energy.

Jules Verne saw the potential of hydrogen as early as 1874, and the question is why this vision has not already been developed. The answer is simple: hydrogen essentially does not occur in a pure form in nature. Energy and a complex technical process are needed before it can be burnt again. This makes hydrogen expensive, and some production processes involved even create high greenhouse emissions. However, with the aid of wind and solar energy, hydrogen, and other gases can be produced in a carbon dioxide-neutral manner.

Jules Verne (1828–1905): 'The Mysterious Island'

"And what will they burn instead of coal?" asked Pencroft. "Water" replied Harding. "I believe that water will one day be employed as fuel, that hydrogen and oxygen which constitute it, used singly or together, will furnish an inexhaustible source of heat and light, of an intensity of which coal is not capable. Water will be the coal of the future."

Today, the production of hydrogen from renewable energies is sometimes referred to as power-to-gas technology. That sounds fascinating, for sure. Technically, the process of producing hydrogen from renewable energies is in fact quite basic. The electrolysis required for this is already known to many from chemistry lessons. The main problem is the use of hydrogen. Large hydrogen storage facilities or transport pipelines are practically non-existent.

Renewable Energy and Climate Change, Second Edition. Volker Quaschning.
© 2020 John Wiley & Sons Ltd. Published 2020 by John Wiley & Sons Ltd.

Step on the Gas

Natural gas is widely used, and many people are, therefore, familiar with it. However, natural gas is a fossil energy source, the combustion of which produces carbon dioxide. Natural gas consists of more than 90% *methane*. In a camping context we are familiar with *propane* or *butane*. Both gases come from natural gas production or petroleum processing and therefore also have a fossil origin.

Before natural gas began its triumphal advance in Germany and other countries, *town gas* was widespread. Town gas is produced from coal and consists only of around 20% methane and 50% hydrogen. Town gas has a lower calorific value than natural gas. For this reason, appliances had to be adapted during the conversion from town gas before they could be connected to natural gas. In West Berlin, the last consumers were not converted to natural gas until 1996.

Biogas consists of only 50–70% methane and also has a lower calorific value than natural gas. If biogas is to be fed into the natural gas grid, it must be conditioned accordingly.

Renewable gases are a new category of gases. Renewable gases are produced with the help of electricity from renewable power plants. Depending on the origin of the electricity, they are also referred to as *wind gas* or *solar gas*. Chemically, it is either *hydrogen* produced from renewable electricity by electrolysis or methane produced from hydrogen in a further process. The main advantage of methane is that it can directly replace natural gas. Combustion of renewably-produced methane also produces carbon dioxide – but only exactly as much as was previously extracted from the atmosphere during gas generation. The use of renewable gases is therefore carbon dioxide-neutral.

Today's gas industry is based on natural gas. In theory, hydrogen can also be added directly to the natural gas network. However, the problem is that the calorific value of hydrogen per cubic metre is significantly lower than that of natural gas. If a gas stove adjusted to natural gas were operated with hydrogen, the performance would decrease rapidly, and the time taken to heat up soup would feel like an eternity. This becomes even more problematic in industrial processes that use natural gas. Uncontrolled admixture of hydrogen could cause major damage. Therefore, the proportion of hydrogen admixture is limited to 2–5%.

In the case of the previously common town gas, around 50% hydrogen was already present in the gas network. Town gas was produced from fossil coal with the help of coal gasification. Since the 1960s, town gas has been successively replaced by natural gas. In Germany, town gas was finally phased out in the 1990s. During the conversion, all the nozzles of consumer appliances had to be readjusted to the changed calorific value. From today's perspective, the conversion is regrettable, as a gas network based on town gas would have been able to handle renewable hydrogen much more easily. A renewed conversion of all consumers to a gas mixture with a higher hydrogen content is not an acceptable alternative due to the high costs involved.

If one wishes to avoid building a hydrogen infrastructure parallel to the natural gas network, a further chemical process can be used to produce methane gas from renewable hydrogen. Since natural gas consists almost entirely of methane, renewable methane can then simply be fed into the natural gas grid. All natural gas storage facilities, pipelines, and consumer appliances can be used directly for renewable methane without conversion. Since Germany has huge natural gas storage facilities, there is also an immediate option to use these facilities for a renewable energy industry. No wonder that power-to-gas technology has become an important source of hope for a carbon-dioxide-free energy supply.

13.1   Hydrogen as an Energy Source

Hydrogen is by far the most common component in our solar system and constitutes around 75% of the mass and more than 90% of all atoms. Our sun and the large gas planets Jupiter, Saturn, Uranus, and Neptune consist primarily of hydrogen (Figure 13.1).

Here on Earth hydrogen occurs much less frequently. Its share of the total weight of the Earth is only about 0.12%. Although hydrogen occurs more frequently in the Earth's crust, it practically never occurs even there as a pure gas. Hydrogen is almost always chemically bonded. The most frequent compound is water.

Hydrogen is the smallest and lightest atom. As an extremely light gas, hydrogen was used to fill the gas bags of airships like the Zeppelins during the first half of the nineteenth century. The Hindenburg disaster, where an electrostatic charge supposedly caused the hydrogen to ignite, brought a tragic end to the prospects of hydrogen use.

The main application of hydrogen today is in the chemical industry. As an energy source it is currently used on a large scale mainly in the aviation sector and in space travel. Hydrogen has occasionally been used to drive the jet engines of aeroplanes. In space travel liquid hydrogen is used as rocket fuel. For example, the launch of a space shuttle consumes about 1.4 million litres of liquid hydrogen weighing more than 100 tons. This is burnt along with the 0.5 million litres of liquid hydrogen that the shuttle carries with it. The combustion temperature is up to 3200 °C.

Hydrogen must first be produced in a pure form before the energy from it can be used. This requires an easily available inexpensive raw material containing hydrogen. Aside

Figure 13.1 Hydrogen is the smallest and lightest atom. Hydrogen was used as energy source for the Space Shuttle. Graphic/Photo: NASA.

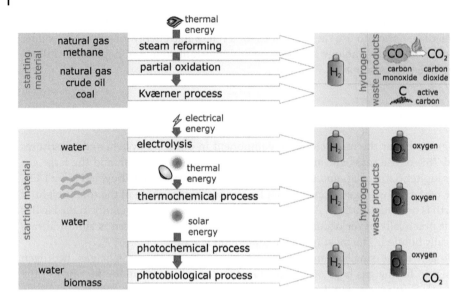

Figure 13.2 Procedures for producing hydrogen.

from water (H_2O), which consists of hydrogen (H) and oxygen (O), hydrocarbon compounds can also be an option. This is primarily natural gas, or methane (CH_4). Heating oil and coal consist of hydrogen (H) and carbon (C) but have a much higher proportion of carbon than natural gas (Figure 13.2).

Current industrial methods for producing hydrogen almost exclusively use fossil fuels, such as natural gas, crude oil, or coal, as the raw material. Methods such as steam reforming or partial oxidation to produce hydrogen from fossil hydrocarbons chemically separate the carbon. It then reacts to form carbon monoxide (CO), which can be utilized. The end product is carbon dioxide (CO_2). These methods for producing hydrogen are therefore not real options for actively protecting the climate.

The Kværner method also uses hydrocarbons as the base material. However, the waste product that it produces is activated carbon, i.e. pure carbon. A direct formation of carbon dioxide can be prevented with this method if the carbon is not burnt further.

Basically, all the methods mentioned to produce hydrogen from fossil energy sources are run at high processing temperatures. This requires large amounts of energy. If this energy comes from fossil sources, this will again lead to the emission of carbon dioxide. For climate protection it is usually better to burn natural gas or oil directly than to take the circuitous route of producing hydrogen and then using it as a supposedly environmentally friendly fuel.

Other methods are therefore necessary to produce hydrogen, so that it is environmentally safe. Electrolysis is the ideal method for this. The German chemist Johann Wilhelm Ritter first used electrolysis to produce hydrogen as early as 1800. Using electric energy, electrolysis decomposes water directly into hydrogen and oxygen. If the energy comes from a renewable energy power plant, the hydrogen can be extracted free of carbon dioxide.

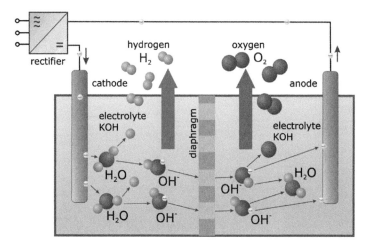

Figure 13.3 Principle of alkaline electrolysis.

Alkaline electrolysis is an example (Figure 13.3). With this method two electrodes are dipped into a conductive watery electrolyte. This can be a mixture of water and sulphuric acid or potassium hydroxide (KOH). The anodes and cathodes conduct direct current into the electrolytes. There they electrolyse water into hydrogen and oxygen.

Although electrolysis has already reached a high state of technical development as an environmentally compatible option for oxygen production, other alternative methods are also in development.

Thermo-chemical methods are an example. At temperatures above 1700 °C water decomposes directly into hydrogen and oxygen. However, these temperatures require expensive heat-resistant facilities. The required temperature can be reduced to below 1000 °C through different coupled chemical reactions. For example, concentrating solar thermal power plants can produce these temperatures, and this has already been successfully proven.

Other procedures include the photochemical and photobiological production of hydrogen. With these procedures, special semiconductors, algae or bioreactors use light to decompose water or hydrocarbons. These methods are also still at the research stage. The main problem is developing long-term stable and reasonably priced facilities.

13.2   Methanation

In order to replace fossil-based gas directly with renewable gas, methane must be produced from hydrogen generated with renewable electricity (Figure 13.4). Methane has the chemical formula CH_4 and consists of four hydrogen atoms and one carbon atom. Carbon dioxide can serve as carbon source. This can come from fossil power plants, biogas plants, or biomass power plants, for example. In principle, it is also possible to extract carbon dioxide from the atmosphere. However, as the concentration here is very low, the carbon dioxide would have to be separated. Various technologies are being developed for this purpose, but these are still energy-intensive and expensive.

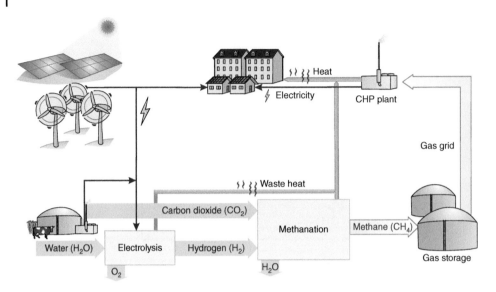

Figure 13.4 Generation, storage, and re-conversion to electricity of renewable methane [Qua13].

Methanation of Hydrogen

The Sabatier process, named after the French chemist Paul Sabatier, converts hydrogen, and carbon dioxide into methane:

$$4H_2 \text{ (hydrogen)} + CO_2 \text{ (carbon dioxide)} \rightarrow CH_4 \text{ (methane)} + 2H_2O \text{ (water)}$$

A catalyst based on nickel or ruthenium is required. During the reaction, heat is released, which can be utilized.

A 25 kW prototype plant that uses carbon dioxide from the atmosphere was built in 2009 at the Centre for Solar Energy and Hydrogen Research Baden-Württemberg. A 250 kW test plant followed in 2012. However, the efficiency of the first plant for the conversion of renewable electricity to methane was still very low, at around 40%. A 6 MW plant with an efficiency of 54% went into operation at Audi AG in 2013. In 2016, a 1.25 MW plant was built by the Haßfurt municipal companies in collaboration with Greenpeace Energy, which converts excess electricity from renewable energies into hydrogen and feeds it into the public gas grid.

13.3   Transport and Storage of Renewable Gas

13.3.1   Transport and Storage of Hydrogen

Once pure hydrogen has been produced and is not to be immediately converted into methane, it has to be stored and transported to the consumer. In principle, we are familiar with the storage and transport of combustible gases from the use of natural gas. Hydrogen is an extremely lightweight gas with very minimal density but has a

Figure 13.5 Experience from the natural gas sector can be used for storing and transporting hydrogen. Left: LNG tanker. Right: Pipeline. Source: BP, www.bp.com.

relatively high calorific value. Compared to natural gas, hydrogen with the same energy content requires much larger storage volumes, although the stored hydrogen is much lighter.

Hydrogen can either be compressed and stored under high pressure or liquefied in order to reduce the necessary storage volumes. Under normal pressure hydrogen condenses but not until it reaches extremely low temperatures of minus 253 °C. Liquid hydrogen is abbreviated as LH_2. A large amount of energy is needed to achieve such low temperatures. Around 20–40% of the energy stored in the hydrogen is used to liquefy it.

In principle, the same technologies used in the natural gas sector can be used for the liquidization, transport, and storage of hydrogen. Hydrogen can be transported either in pipelines or in special tankers and freighters. Whereas pipelines usually transport the gaseous form, tankers are preferred for liquid hydrogen to reduce the volume. In contrast to hydrogen, natural gas already becomes liquid at minus 162 °C and is abbreviated as LNG (liquefied natural gas) (Figure 13.5).

Compressed gas or liquid gas tanks are used to store small quantities. Another disadvantage of hydrogen is the very small atomic portion, which makes it extremely volatile. Large quantities can be lost if it is stored in metal tanks for long periods of time, because it diffuses through the storage walls. Underground storage facilities similar to those already used for natural gas are also suitable for storing large quantities of hydrogen. Hydrogen is injected into underground cavities at high pressure and can be removed again as required.

13.3.2 Transport and Storage of Renewable Methane

Renewable methane can be fed directly into the natural gas grid. This means that the existing natural gas infrastructure can be used without modification. Transport over long distances takes place through high-pressure pipelines with a diameter of 1–2 m. Compressor stations located every 100–200 km maintain the high pressure of up to 100 bar. Smaller medium- and low-pressure supply lines distribute the gas locally to

Figure 13.6 Underground natural gas storage facilities in Germany. Source: Map of Germany: NordNordWest, www.wikipedia.de, Data on storage facilities: LBEG Lower Saxony D [LBEG12].

the end users. The pressure at the consumer appliances is only around 20 mbar. Gas pipelines with a total length of over 400 000 km have been laid in Germany.

Large quantities of natural gas are stored underground (Figure 13.6). A distinction is made between pore or adsorptive storage of natural gas and cavern storage. Depleted natural gas or oil reservoirs are already being used as pore storage facilities. Pores and fissures of underground layers of lime and sandstone absorb the gas. The geological condition of pore reservoirs is generally well-known and the tightness has been proven by the natural gas or oil originally stored for millions of years.

Table 13.1 Capacities for the storage of hydrogen and methane in Germany [LBEG16]

	Working gas volume in billion m³	Storage capacity in billion kWh for	
		Hydrogen	Methane
20 pore storage facilities in operation	9.8	—	98
31 cavern storage facilities in operation	14.5	44	145
7 cavern storage facilities in planning or construction	4.0	12	40
58 storage facilities in operation, planning, or construction	28.3	56	283

For cavern storage facilities, artificial caverns are created in salt domes. Salt is washed out, through a well, with water at a depth of several hundred metres, creating large cavities of up to 100 million cubic metres. The underground caverns reach heights of up to 500 m. Several individual caverns can be combined to form an even larger cavern storage facility. Due to the salt above the cavity, the reservoir is naturally tight and can only be filled and discharged through the existing boreholes. Not all of the gas can be taken out again. About one third of the cavity must remain filled with so-called cushion gas to ensure the pressure and stability of the reservoir. The remaining storage volume can be continuously charged and discharged with the working gas.

In total, Germany has 51 pore and cavern storage facilities with a working gas volume of 24.3 billion cubic metres. A further 7 storage facilities with a storage volume of 4.0 billion cubic metres are being planned or under construction (Table 13.1). Cavern storage facilities can also be used to store hydrogen. All existing or planned cavern storage facilities could then absorb hydrogen with an energy content of 56 billion kilowatt hours. With methane, the storage capacity is even greater. The total storage potential amounts to 283 billion kilowatt hours.

Total natural gas consumption in Germany in 2016 was around 850 billion kilowatt hours. This would allow the storage facilities to cover Germany's current natural gas demand for around three months. Gas demand could increase even further if, in future, gas storage facilities are to be used increasingly for the long-term storage of surpluses from solar and wind power plants. But even then, with the existing and planned gas storage facilities, there is sufficient capacity to guarantee a secure supply based on renewable energies.

13.4 Fuel Cells: Bearers of Hope

Fuel cells are considered a key technology for the future energy use of hydrogen, because they can convert hydrogen directly into electric energy. Theoretically at least, this results in higher efficiency levels than with combustion in conventional thermal power plants.

The principle of fuel cells has been known for a very long time. There is some controversy about who actually invented the technology. The German-Swiss chemist

Christian Friedrich Schönbein conducted the first tests in fuel cell technology in 1838. The English physicist Sir William Robert Grove built the first fuel cell in 1839. Well-known scientists like Henri Becquerel and Thomas Edison were subsequently involved in its further development. A sufficiently advanced stage of development was finally reached in the mid-twentieth century, enabling NASA to make major use of fuel cells by 1963.

Since the 1990s, fuel cell development has been moving ahead at full speed. Car manufacturers and heating companies have adopted the technology and are looking to profit from a positive image as a result.

Fuel cells basically involve a reversal of electrolysis. A fuel cell always contains two electrodes. Depending on the type of fuel cell, pure hydrogen (H_2) or a fuel containing hydrocarbons, is fed through the anode and pure oxygen (O_2) or air as an oxidation material, is fed through the cathode. An electrolyte separates the anode and cathode (Figure 13.7). As a result of this, the chemical reaction is controlled. Electrons flow over a large circuit and emit electric energy. The remaining positively charged ions diffuse through the electrolyte. The waste product is water.

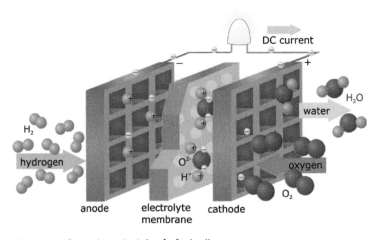

Figure 13.7 Operating principle of a fuel cell.

There are different types of fuel cells that essentially differ from each other based on electrolytes, the permissible fuel gases and operating temperatures. In practice, the following abbreviations are used to identify the fuel cell types:

AFC	alkaline fuel cell
PEFC	polymer electrolyte fuel cell
PEMFC	proton exchange membrane fuel cell
DMFC	direct methanol fuel cell
PAFC	phosphoric acid fuel cell
MCFC	molten carbonate fuel cell
SOFC	solid oxide fuel cell

Figure 13.8 shows the respective fuel gases and oxidation materials as well as the electrolytes and operating temperature ranges for the different types of fuel cells.

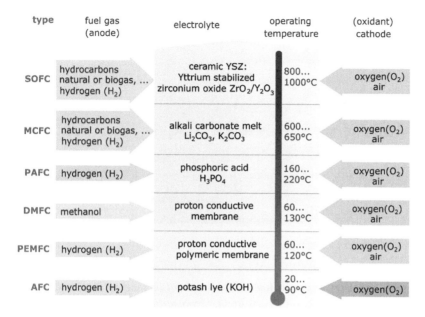

type	fuel gas (anode)	electrolyte	operating temperature	(oxidant) cathode
SOFC	hydrocarbons natural or biogas, ... hydrogen (H_2)	ceramic YSZ: Yttrium stabilized zirconium oxide ZrO_2/Y_2O_3	800... 1000°C	oxygen(O_2) air
MCFC	hydrocarbons natural or biogas, ... hydrogen (H_2)	alkali carbonate melt Li_2CO_3, K_2CO_3	600... 650°C	oxygen(O_2) air
PAFC	hydrogen (H_2)	phosphoric acid H_3PO_4	160... 220°C	oxygen(O_2) air
DMFC	methanol	proton conductive membrane	60... 130°C	oxygen(O_2) air
PEMFC	hydrogen (H_2)	proton conductive polymeric membrane	60... 120°C	oxygen(O_2) air
AFC	hydrogen (H_2)	potash lye (KOH)	20... 90°C	oxygen(O_2)

Figure 13.8 Differences between fuel cell types.

The PEMFC is the one most frequently used today. In this fuel cell the electrolyte consists of a proton-conductive polymer film. The fuel gases flow through carbon or metal substrates, which serve as electrodes. The substrates have a platinum coating that acts as the catalyst. The typical operating temperature is about 80 °C. These cells do not require pure oxygen for operation but can also work with normal air.

Because hydrogen as an energy source is only available in limited quantities today, there is an interest in using fuel cells directly with energy sources like natural gas and methanol that are relatively easily available. At a preliminary stage a reformer uses a chemical process to break down hydrocarbons such as natural gas into hydrogen and other components. This process creates a hydrogen-rich reformate gas, after which a gas purification stage is still needed to eliminate harmful carbon monoxide (CO) for the fuel cell.

MCFC and SOFC work at much higher temperatures. This enables gases containing hydrocarbons, such as natural gas and biogas, to be used directly without any previous reforming. The disadvantage of high temperatures is the long time required for starting up and switching off.

As the electric voltage of a single cell with values of around 1 Volt is too low for most applications, a number of cells are usually connected in series to form a so-called stack (Figure 13.9).

The electric efficiency of fuel cells today is usually in the order of 40–60%. Values of more than 60% can be reached in individual cases. A fuel cell can therefore only convert a part of the energy contained in hydrogen into electric energy. Basically, the waste heat of fuel cells is also usable. Power-heat coupling, or the simultaneous generation of power and heat, raises the overall efficiency of a fuel cell and can increase it to over 80%.

Figure 13.9 Fuel cell prototypes.

Major advances have been made in fuel cell technology in recent years, and many companies are already offering commercial units. However, the number of units currently being sold is still relatively low. The price of fuel cell systems is still fairly high compared to other energy supply units. Furthermore, it will be necessary to increase the sometimes quite short service life of fuel cells if they are to have a broader appeal.

13.5 Economics

It currently costs about 4 cents to produce a kilowatt hour of hydrogen through the steam reforming of natural gas – assuming that natural gas prices are relatively reasonable. One litre of petrol has a calorific value of about 10 kW h. Added to this would be the equivalent of hydrogen to 1 l of conventional petrol at about 40 cents. At this point the hydrogen would not yet even have reached the tank of the consumer. Including liquidization, transport, and storage, the cost rises threefold from the given price to well over €1. This makes it more than double the net petrol price in Europe in 2017.

Producing hydrogen in an environmentally compatible way through electrolysis using renewable energy is even more expensive. With a prototype facility using wind power for electrolysis, the equivalent of hydrogen to 1 l of conventional petrol would cost €5. With large-scale technical plants, a price of €2 could easily be achieved at the refuelling pump. Taxes and duties would have to be added to this.

The medium-term hope is that hydrogen at prime renewable sites using electricity from wind turbines, hydropower plants, or solar power plants will be available for delivery at the equivalent of less than €2 l^{-1} of petrol. If the petrol price then rises well above €2 l^{-1}, hydrogen at the pump would become competitive. Presumably it will be some time before we get to that stage.

In addition to its use as a fuel, hydrogen is also considered an option for the large-scale storage of electrical energy. Relatively large losses occur during the electrolysis, storage, and re-conversion to electricity of hydrogen or methane (Figure 13.10). Therefore, this only makes sense if the electricity is available at a very reasonable price, for example due

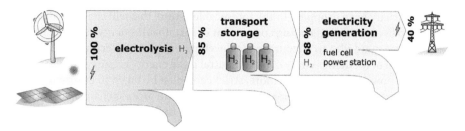

Figure 13.10 Losses when hydrogen is used to store electric energy, based on the current state of the art.

to a very high supply of solar and wind power at times. The use of waste heat in the production of hydrogen or methane and in power recovery could significantly improve the economic efficiency. From today's perspective, some technological progress and, above all, an expansion of combined heat and power generation are still necessary to make the use of renewable hydrogen or methane in the electricity industry meaningful from an economic point of view. As seasonal storage will not yet be necessary on a large scale within the next 10 years, it can be assumed that the necessary progress may still be made in good time.

13.6   Ecology

The broad public perception of hydrogen as an energy source and fuel cells is favourable, mainly because water is the waste product when hydrogen is used.

However, what is important for the environmental balance is not what comes out at the end but what is put in at the beginning. When steam reforming is used to produce hydrogen from natural gas, around 300 g of carbon dioxide is created per kilowatt hour of hydrogen (g CO_2/kWhH$_2$); with partial oxidation of heavy oil this rises to even 400 g CO_2/kWhH$_2$ [Dre01]. This is clearly more than is created through the direct use of natural gas and crude oil. If hydrogen is extracted by electrolysis using average electricity in countries like Britain or Germany, the carbon dioxide figure rises to more than 600 g CO_2/kWhH$_2$. Hydrogen as an energy source then ends up being far from ideal for the environment. For the greenhouse gas balance, it ultimately makes more sense to continue driving petrol and natural gas-powered cars than to change them over to run on hydrogen.

Hydrogen only offers a true alternative if it is extracted using pure renewable energy sources, such as through electrolysis using energy from wind turbines and solar power plants. However, as long as hydrogen is produced using methods that can create carbon dioxide, it will, at best, be suitable for testing prototypes.

Many manufacturers are developing products that allegedly rely on clean hydrogen as an energy source and fuel cells for generating electricity. They owe us an explanation of how they intend to make sufficiently large quantities of reasonably priced carbon-free hydrogen available soon.

In recent years, prices for solar and wind power plants have fallen significantly and the trend is continuing. As a result, the production costs for hydrogen produced with renewable electricity have dropped. In the foreseeable future, this will enable hydrogen and fuel cells to become an interesting component of sustainable energy supply. Therefore, it is already a good idea to encourage the development of these technologies.

The production of methane is only climate-neutral if the energy required for it comes from renewable power plants. Furthermore, strict care must be taken to ensure that no unused methane escapes into the atmosphere along the entire process chain, as methane has a significantly higher global warming potential than carbon dioxide. Here the risk of direct use of hydrogen is considerably lower.

13.7   Markets, Outlook, and Development Potential

Hydrogen production is currently at a total of 45 million tons worldwide. In comparison, crude oil consumption at 4418 million tons worldwide in 2016 was higher by several orders of magnitude. As the chemical industry uses the bulk of the available hydrogen, a market for it does not yet exist in the energy sector. The capacity for hydrogen production using renewable energies is very low, and there is also no infrastructure for the transport and storage of large volumes of hydrogen.

Only a small number of hydrogen refuelling stations currently exist to supply hydrogen for filling the tanks of hydrogen test vehicles. The cost of developing a comprehensive network of stations that could provide hydrogen fuel is estimated in the billions of euros in countries like Germany and Britain alone. In larger countries such as the USA this sum would be even higher. As long as hydrogen is still much more expensive than conventional fuels, the chances of this kind of investment are very low (Figure 13.11).

Whereas electric drives with batteries charged using renewable energies offer an alternative for cars, climate-neutral concepts are still lacking for the powering of aeroplanes. Hydrogen could be the solution in this area. However, as planes operating on hydrogen also emit water vapour and produce condensation trails, which in turn contributes to the greenhouse effect, air traffic based on hydrogen would not be totally climate-neutral.

In recent years, a few prototype plants have been built for the production of renewable methane. However, the quantities of methane that can be produced are so small that they are insignificant from an energy economy perspective.

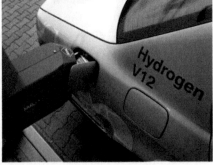

Figure 13.11 Many car makers and energy companies are banking on hydrogen. However, the network of hydrogen fuel stations is still extremely small and is only capable of filling up a few prototypes that use hydrogen. Photos: BP, www.bp.com.

If the share of solar and wind energy in the electricity supply increases noticeably to more than 50%, the importance of long-term storage facilities will also increase. Although renewable hydrogen or methane is a very promising option, it is expected to take at least another 10 years before it can be used on a larger scale.

Technical developments in fuel cell technology are progressing at a much slower pace than the product announcements in the 1990s may have led us to believe. A few commercial projects exist at the moment, but the number of units sold is still relatively low. Technical advances are also needed in the fuel cell arena before a larger market can be exploited. Above all, a functioning renewable hydrogen or methane industry is needed before fuel cells can become a truly climate-compatible alternative.

14

Sunny Prospects – Examples of Sustainable Energy Supply

The previous chapters covered the large spectrum of renewable energies available. Renewable energies will largely have to replace conventional energy sources during the next 20 years to keep the consequences of climate change within manageable limits. The modern use of these energies does not have decades of tradition to fall back upon. Their applications are often new and unusual.

Yet climate-compatible energy supply is not pie in the sky. A life beyond fossil and nuclear energy sources is already possible today. This chapter illustrates this based on impressive examples from different areas.

14.1   Climate-Compatible Living

Despite the pressing need to protect the climate, low-energy dwellings such as 'three-litre' houses or passive houses meeting the German KfW 55 or KfW 40 standards are still the exception rather than the rule. Most buildings that undergo restoration or renovation also fall short of meeting the standard for climate compatibility.

It is technically as well as economically easy to incorporate energy supplies that are low or neutral in carbon dioxide into new buildings in particular. It is far simpler to do this at the building stage rather than adapting the finished house in the future. Energy-saving adaptations are usually postponed until windows, roofs, and outside walls are due for refurbishment anyway. However, with well-constructed new buildings this can take many years. If energy prices continue to rise over the next few years, then even new-builds with unnecessarily high requirements in energy will become very costly to run. Investing in high quality insulation and renewable energy supply when a house is being built or renovated means not having to worry about increases in energy prices in the years to come.

14.1.1   Carbon-Neutral Standard Prefabricated Houses

Most of the climate-protecting measures available in building construction are relatively unspectacular. The first example shows an ordinary single-family home in Berlin with about 150 m^2 of living space. It was built in 2005 using timber frame construction. Compared to solid construction, wood frame construction offers good insulation values even with thinner walls. In this particular house, the wall thickness was increased to 32 cm by adding 8 cm of insulation on the outside, to reduce further heat losses through the walls.

Renewable Energy and Climate Change, Second Edition. Volker Quaschning.
© 2020 John Wiley & Sons Ltd. Published 2020 by John Wiley & Sons Ltd.

Figure 14.1 Left: Family home in Berlin with carbon-neutral energy supply. Right: Boiler room with buffer storage tank, solar cycle pump, controlled heat recovery system and wood pellet boiler.

The standard thermal windows were replaced using triple-glazed windows with a high insulation value. A controlled ventilation system with heat recovery reduces ventilation losses and increases comfort levels. These measures alone will enable this building to reach the standards of a 3-l house (Figure 14.1).

A wood pellet boiler and a $4.8\,\mathrm{m}^2$ solar thermal system cover the remaining space heating and hot water demand. A pellet store $5\,\mathrm{m}^2$ in size has capacity for around 6 tonnes of wood pellets. This covers heat energy demand for around three years and is even enough to bridge periods when fuel prices are particularly high. A photovoltaic (PV) system with a capacity of 1 kW peak, which was later expanded to 10.7 kW peak, generates around twice as much electricity per year as is consumed in the house.

The additional costs for the measures described compared to the minimum newbuild standard in 2005 amounted to around €30 000. Retrofitting of the PV system added another €15 000. The cost of wood pellets is around €500 yr^{-1}. The payments for the solar electricity fed into the grid are significantly higher, so that overall the space heating and electricity consumption costs for this house are virtually zero.

A green power supplier covers the electricity demand at times of low sunshine in a carbon dioxide-free manner. Since the PV system generates significantly more electricity than is consumed in the house, the house is not only carbon dioxide-neutral during operation, it even provides an additional reduction in greenhouse gases.

14.1.2 Plus-Energy Solar House

If the entire roof area of an optimally insulated house is fitted with solar systems, a carbon dioxide-neutral house can be turned into a plus-energy solar house. An example is the house shown in Figure 14.2, located in Fellbach near Stuttgart. Optimal heat insulation using cellulose, triple-glass solar glazing, a ventilation system with heat recovery and preheated air supplied from the ground reduce the heat energy needs by about 80% compared to a standard new-build.

A brine-water heat pump with an electrical capacity of 1.1 kW covers the remaining space heating demand of the house. The low-temperature heat comes from two

Figure 14.2 Plus-energy solar house in Fellbach near Stuttgart. The PV system feeds more electricity into the public grid each year than the house needs to cover its own space heating and electricity demand. Source: Reinhard Malz, www.fellbach-solar.de.

vertical ground probes sunk 40 m into the ground. A 0.50 kW heat pump would have been sufficient but was not available on the market.

The PV system with a capacity of 8 kW peak feeds more than 8000 kWh into the public grid each year. This is more than the electricity demand of the house, including the ventilation system and the heat pump. As a result, the energy supplied in this house is not only carbon-neutral but actually saves carbon dioxide elsewhere due to the solar electricity fed into the grid.

14.1.3   Plus-Energy Housing Estate

Houses that produce an energy surplus over the year and feed it into the grid do not have to be an exception. This is demonstrated by the Schlierberg solar housing estate in Freiburg in southern Germany. The architect Rolf Disch designed 50 plus-energy houses for this estate. He has been integrating ecological enhancements and renewable energies into his projects for more than 30 years.

All houses in this development of terraced houses are orientated towards the south. The gaps between the terraces are designed so that the heat from the sun can stream into the houses through large south-facing windows in the winter. Balconies that jut out prevent the houses from heating up too much in the summer. These simple and cheap measures could reduce the energy needs of many new-build houses.

Optimal insulation that exceeds the standard and controlled living space ventilation with heat recovery are standard features of this particular housing estate. These features are designed to keep heat energy needs to an absolute minimum (Figure 14.3).

A woodchip combined heat and power (CHP) plant covers the rest of the remaining space heating demand and provides a climate-neutral energy supply. A local heating network transports the heat to the individual houses. The PV systems feed into the public electricity grid. The solar energy surplus produced by the houses makes them plus-energy homes.

Figure 14.3 Schlierberg solar housing estate in Freiburg, Germany, with 50 plus-energy houses. Source: Rolf Disch SolarArchitektur, www.solarsiedlung.de.

The building materials were also chosen with ecology and sustainability in mind. The wood has not been chemically treated, solvent-free paints and varnishes have been used throughout and the water pipes and electric cables are PVC-free.

14.1.4 Heating Only with the Sun

Many houses use solar thermal systems only to heat water in the summer and at best to provide heating support during the transitional seasons. It is hard to imagine that houses in countries with cold winters could be heated with solar energy alone.

But in recent years many single-family houses have proven that the heating demand can be covered by the sun even in Central Europe. The first multiple-family home in Europe to be heated 100% with solar thermal heat was built in Burgdorf near Bern, Switzerland, in 2007 (Figure 14.4).

Figure 14.4 Left: First multiple-family home in Europe with 100% solar heating. Right: Installation of 276 m² roof-integrated solar collector. Photos: Jenni Energietechnik AG, www.jenni.ch.

High-quality heat insulation and a ventilation system with heat recovery ensure that heat losses in this building are relatively low. The large 276 m^2 roof is completely covered with solar collectors. An enormous hot water storage tank with a capacity of 205 000 l is located in the centre of the building. The seasonal storage tank extends from the basement to the loft. The collectors heat the storage tank with solar heat in the summer. In winter, when the collectors cannot supply sufficient heat, the heat from the storage keeps the building cosy. No auxiliary heating is required.

Less than 10% of the total construction cost of around 1.8 million euros for the building was spent on the solar system. Yet aside from the small expenditure for electricity for the solar and heating circuit pumps, no additional heating-related costs are incurred. Heating costs are included in the rent, so that residents of the block do not have to worry about constant rises in energy bills. Since the house exceeded all expectations during actual operation, two further solar-heated multi-family houses were built on the same site.

14.1.5   Zero Heating Costs After Redevelopment

Practically all technical options can be incorporated into new-builds without a problem. The situation is considerably more complex with existing buildings that date from a time before protecting the environment was considered important.

Building renovations or the installation of a new heating system present the ideal opportunity to make enormous reductions in carbon dioxide. One example is the zero-heating cost apartment block in Ludwigshafen, Germany, which was built in the 1970s and then renovated in 2007 to a technical standard comparable to today's levels (Figure 14.5).

As a result, space heating demand is reduced by around 80%, which now amounts to only around 20 kWh m^{-2} of living space per year. 30 cm thick outer wall insulation and triple-glazed thermal windows provide optimal building insulation. Controlled living space ventilation with heat recovery reduces ventilation heat losses. Solar thermal façade collectors provide for the hot water supply and a large grid-coupled PV system is located on the roof. The minimal residual space heating demand is covered electrically. The yield

Figure 14.5   Constant hikes in energy bills are a thing of the past for this zero heating-cost house in Ludwigshafen, Germany. Photos: LUWOGE, www.luwoge.de.

from the PV system should be roughly the same as the heating electricity costs. As a result, the tenants have zero heating costs. Constant rises in heating costs are therefore a thing of the past for this building. Taking into account all costs including upkeep and vacancy rates, the modernization measures have also turned out to be very lucrative for the investor. Both sides benefited from the climate-protection measures that were taken.

14.2 Working and Producing in a Climate-friendly Manner

14.2.1 Offices and Shops in the 'Sonnenschiff'

The 'Sonnenschiff' (Sun Ship) is located near the housing estate in Freiburg presented above. The Sonnenschiff is a solar service centre that houses shops, offices, medical practices, and penthouse flats. The plus-energy design for the building complex was developed by the architect Rolf Disch. Optimal insulation incorporating advanced vacuum insulation material and intelligent ventilation with heat recovery help reduce the heating demand. In winter the sun streams through large windows right into the heart of the building, providing some of the heating. In summer canopies and blinds prevent the building from overheating (Figure 14.6).

A woodchip CHP plant covers the small residual heating demand. PV systems are not simply mounted on the roof but are actually used instead of a conventional roof. As a result, the entire roof surface generates electricity.

An intelligent solution has also been devised for transport connections. The complex has its own tram stop and conveniently sited parking areas for bicycles, thereby offering alternatives to motor car use. The electric cars of a car-sharing firm are allocated preferential parking spaces and can be charged using solar power generated by the complex itself. The Sonnenschiff complex has put into practice the concept of working and living carbon-free and contributing to climate-compatible mobility.

14.2.2 Zero-Emissions Factory

It is also relatively easy for factories to operate without emitting carbon dioxide. An example is the Solvis zero-emissions factory in Braunschweig, Germany, which produces solar and pellet systems. High-quality heat insulation and waste air recovery in the factory have resulted in energy savings as high as 70% compared to conventional

Figure 14.6 The 'Sonnenschiff' (Sun Ship) in Freiburg contains two shops, a bistro, offices, and medical practices. Photo: Rolf Disch SolarArchitektur, www.solarsiedlung.de.

Figure 14.7 Solvis zero-emissions factory in Braunschweig, Germany. Photo: SOLVIS, www.solvis.de.

structures. Optimal use of daylight and low-energy office machines have halved electricity needs. Solar collectors and PV systems generate 30% of the electricity and heating required. A rapeseed oil CHP plant covers the remaining energy needs, thereby ensuring that all energy supply is completely carbon-free.

Large amounts of heat often escape through the open shed doors of factories during the loading and unloading of heavy goods vehicles. This factory keeps heat losses at a minimum because the loading and unloading zones are situated inside the building. A direct connection to local public transport, parking facilities for bicycles, and showers for cyclists provide employees with a climate-compatible workplace (Figure 14.7).

14.2.3 Carbon-free Heavy Equipment Factory

The 'factory of the future', completed in May 2000 by Wasserkraft Volk AG, is the first energy self-sufficient carbon-free heavy equipment factory built in Germany. The administration building is mainly built of timber from the Black Forest area. The entire building including the factory floor is very well insulated. The offices are orientated towards the south, and the factory has large roof lights.

This construction enables maximum use of daylight and the incoming sunlight covers some of the heating demand. A 30 m^2 solar collector on the administration building actively utilizes solar energy. At the heart of the factory's energy supply is its own hydropower plant with a capacity of 320 kW. The turbines integrated into the building use natural water from the Elz River nearby. The waste heat from the generator covers about 10% of the heat demand. Three heat pumps with a heat output of 130 kW extract heat from the groundwater and cover the remaining heat demand.

The hydropower plant generates an annual surplus of around 900 000 kW h of electric energy, which the factory feeds into the public grid. This makes it not only a carbon-free factory, but also a plus-energy one that supplies additional green electricity (Figure 14.8).

14.2.4 Plus-Energy Head Office

The buildings of Juwi AG in Wörrstadt, Rhine-Hesse, prove that even large company headquarters can be completely supplied by renewable energies. The company complex was commissioned in 2008 and subsequently extended several times.

Figure 14.8 The administration building at the Wasserkraft Volk AG factory of the future in Gutach, Germany, has its own hydropower plant, which generates carbon-free electricity. Photos: WKV AG, www.wkv-ag.de.

The buildings mainly use timber as an environment-friendly building material. Optimum insulation and a sophisticated ventilation system with heat recovery reduce the energy demand of the buildings. A roof-mounted solar thermal system, two wood pellet boilers with a total capacity of 590 kW, a biogas cogeneration plant, and several ground heat wells cover the heating demand in a carbon dioxide-neutral manner. Cooling is supplied by an absorption chiller (Figure 14.9).

When purchasing the electrical equipment, attention was always paid to maximum efficiency, so that the electrical energy demand could also be considerably reduced. Several PV systems integrated into the building and carports with solar roofs cover a large proportion of the electricity demand. Together with the biogas CHP plant, the company complex generates more electricity than it consumes.

The company's own shuttle service and numerous electric vehicles in the company's fleet reduce transport-related carbon dioxide emissions. The company's own canteen

Figure 14.9 Energy-efficient, purely renewably supplied company headquarters of Juwi AG in Wörrstadt with building-integrated PV system and battery storage. Photos: Mosler for JUWI, www.juwi.de.

relies on regional organically grown food and thus rounds off the company's sustainability ethos.

14.3   Climate-Compatible Driving

The colourful, glossy brochures produced by leading car manufacturers are full of impressive assertions about how much they are doing to protect the environment. However, with many manufacturers it is difficult to discern any real innovation in creating carbon-free means of transport. Cars that run on natural gas and hybrid cars make only a marginal difference in reducing carbon dioxide emissions, and electric cars that can be charged with green electricity are only slowly entering the market. The sustainable cultivation potential with biomass fuels is too low to meet total needs and there is simply not enough carbon-free hydrogen being produced to advance the innovation of hydrogen cars. For a long time, the US pioneer Tesla was the only one to launch purely electric cars with a range suitable for everyday use in the luxury segment, but now other manufacturers are also following suit.

In the short term, only plug-in hybrids or pure electric cars can offer an alternative for climate protection. Plug-in hybrid vehicles are normally recharged through a power point and for emergencies have a combustion engine that can run on biofuel. Pure electric cars only run on electricity. If the electricity is supplied by renewable power plants, no carbon dioxide is emitted while the vehicle is being driven. The potential of electric cars is demonstrated by some impressive examples.

14.3.1   Travelling Around the World in a Solar Car

Solar cars have a reputation of being only suitable for short distances. They are usually not trusted to be driven for long distances, and under challenging conditions (Figure 14.10).

To highlight the problems of climate change, Louis Palmer of Switzerland set off on the first trip around the world in a solar-power taxi on 3 July 2007. Palmer had been

Figure 14.10 In July 2007 Louis Palmer of Switzerland began his trip around the world in a solar energy-operated vehicle. Photos: www.solartaxi.com.

dreaming about a solar taxi since 1986. In 2005 he finally succeeded in implementing the project, thanks to the help of numerous sponsors and technical support from the Swiss Federal Institute of Technology (ETH) in Zurich.

His route took him more than 50 000 km through 50 countries on 5 continents over 15 months. On the trip he and his team took advantage of numerous stops and events to introduce solar technology and provide impetus for the use of new climate-friendly technologies. Allowing people to test drive it enthused many people in the countries it toured.

The solar taxi is designed as an electric car. It has a 5-m trailer that is covered with 6 m^2 of solar cells. A new type of battery stores the electricity, thereby enabling the car to be driven at night and when there is no sun. The daily range is about 100 km.

14.3.2 Across Australia in 33 hours

The World Solar Challenge proves that solar cars are already capable of extremely high performance. This event has taken place regularly in Australia since 1987 and is the most demanding solar car race in the world. The race follows a route on public roads for around 3000 km right across Australia: from Darwin in the north, to Adelaide on the south coast. The racing teams try to cover maximum distances between the hours of eight in the morning and five in the afternoon (Figure 14.11).

The vehicles battling for first place in the race are technical masterpieces. The size of the batteries is restricted by the rules and the only energy allowed is what is supplied by solar modules that are mounted directly onto the vehicles. Powerful solar cells with efficiencies of well over 20% provide the necessary drive energy. The cars are optimized for aerodynamics and trimmed to a minimal weight. As a result of constant technical advances and more and more efficient solar cells, the average speeds of the winners have been increasing steadily. In 2005 it was 102.7 km h^{-1}. As Australia imposes speed limits on public roads, it will not be possible from a practical standpoint to increase this speed much further. Therefore, in 2007 the rules were changed to limit the size of the solar generator and stipulate a seated position for the driver.

Figure 14.11 In 2007 the solar car from Bochum University reached an average speed of 73 km h^{-1} during the 3000 km race. Photos: Bochum University of Applied Sciences, www.hs-bochum.de/solarcar.html.

Teams from all over the world have been participating in the competition for many years. In 2007 the Solarworld No. 1 racing car from Bochum University of Applied Sciences reached fourth place. It took the team from Bochum around 41 hours of pure driving time to cover the 3000 km distance.

In 2012, after the World Solar Challenge the Bochum team circumnavigated the globe in its optimized solar car, like Louis Palmer before, covering almost 30 000 km. Thanks to optimized solar technology, they no longer needed a solar trailer or a power socket.

14.3.3   Emission-free Deliveries

If you think about it, there is nothing less suitable for inner-city delivery traffic than a vehicle with a diesel engine. It does not like frequent starts and stops, contributes significantly to air pollution in cities and also causes a high level of noise pollution. With daily journeys of around 100 km, end-customer delivery traffic is therefore predestined for the use of electric vehicles. That's what Deutsche Post thought, too. But it could not find a manufacturer that could or wanted to supply suitable vehicles.

As a result, and without further ado, Deutsche Post bought StreetScooter GmbH in December 2014. The company had previously developed a prototype for an electrically powered van as a privately organized research initiative at RWTH Aachen University. In 2016, StreetScooter started series production and expanded its annual production capacity to 20 000 vehicles by 2018.

The basic version has a service load of 650 kg and achieves a nominal range of 118 km and a maximum speed of 85 km h^{-1} with a 20 kWh lithium-ion battery. StreetScooter now also cooperates with conventional car manufacturers in the marketing of vehicles and the development of larger vans (Figure 14.12).

The major car manufacturers have now realized that they need to review their lack of commitment to clean mobility, having lost much credibility with their conventional business model.

Figure 14.12   Environment-friendly electric vehicles are predestined for inner-city delivery traffic. Photos: StreetScooter GmbH, www.streetscooter.eu.

14.3.4 Electric Cars for All

The Munich-based startup Sono Motors is a good example for understanding why conventional car manufacturers have tried to avoid electric mobility for a long time, preferring instead to adhere to the internal combustion engine by hook or by crook, sometimes by unfair means. At the beginning of 2016, three friends with a vision founded the company because they had grown impatient with the slow pace of the departure from oil in the transport sector. Through crowdfunding, they collected the necessary money to build a prototype called Sion, which they presented to the public in 2017. They were able to inspire many test drivers with the vehicle. The car has a nominal range of 320 km and, among many other technical refinements, has a PV system integrated into the body. This allows the car to gain up to 30 km d^{-1} in additional range directly from the sun, without having to connect to a power outlet. The estimated retail price is €16 000 plus the cost of the battery, which is likely to add another €4000–6000.

Whether the company will be successful in the long term remains to be seen. But the mere fact that three newcomers succeeded in designing a fully-functional, production-ready car is impressive. The essential components for such a car are freely available to everyone on the supply market; plus, compared to complex petrol or diesel engine technology, the integration of batteries and electric motors is simple. So, if an electric car can in principle be developed and built by anyone, it poses enormous problems for the major car manufacturers. With the combustion engine they lose their main know-how and their unique selling point (Figure 14.13).

Since electric cars require considerably less maintenance, have far fewer wearing parts and don't need oil changes, the lucrative car repair shop business also collapses. No wonder then, that the car companies have, for many years, avoided the electric car like the plague. But they can no longer stop its triumphal advance. Electric cars will soon be cheaper than comparable cars with combustion engines. They are more environment-friendly and can actually be more fun to drive. The smartphone has shown us what rapid upheavals await us in the automotive sector. In a few years' time we will ask ourselves why we drove vehicles that relied on combustion for so long. If we accelerate the

Figure 14.13 Electric cars are much easier to build than cars with combustion engines. Photos: Sono Motors, www.sonomotors.com.

expansion of electricity generation from renewables at the same time, it is also an opportunity for the rapid development of climate-neutral road transport.

14.4   Climate-Compatible Travel by Water or Air

14.4.1   Advanced Sailing

As a result of globalization, goods are being shipped over ever-longer distances. A large part of the increase in transport is being handled by commercial shipping. The carbon dioxide emissions per transport kilometre are significantly lower for shipping compared to air freight. Nevertheless, shipping is also contributing noticeably to the greenhouse effect. Around 3% of all global carbon dioxide emissions are attributed to shipping – and the rate is rising.

Until the middle of the nineteenth century, sailing ships dominated freight and passenger traffic at sea. Steamships then came along and had the advantage that they did not have to depend on wind conditions to keep to their timetables. They gradually replaced sailing ships, which today are used almost exclusively for leisure and sporting activities.

There are various new concepts that make wind power available in combination with conventional ship propulsion systems. The German inventor Anton Flettner developed a cylindrical rotor to propel ships in the 1920s. However, this propulsion method did not catch on at the time. New ship prototypes are currently in development, including the Flettner rotor combined with a conventional ship diesel drive, aimed at reducing fuel requirements from 30% to 40%.

Another interesting concept is based on modern towing kites (Figure 14.14). A starting and landing system automatically lowers and raises the power kite and the cabling rope. The kite flies at a height between 100 and 300 m. The wind at this height is stronger and more consistent than on the deck of a conventional sailing ship. The towing cable,

Figure 14.14   Advanced, automatically controlled towing kites can reduce the fuel consumption of conventional ships by up to 50% and, when combined with biofuels, even make it CO_2-free. Photos: SkySails, www.skysails.de.

which is made of modern synthetic material, is attached to the foreship, and the towing force is transferred through to the ship. The steering of the kite is completely automated. Through a shortening or lengthening of the steering lines, a power kite can be controlled like a paraglider and orientated with optimal precision depending on wind direction, wind intensity and a ship's course. With a sail surface of up to $5000\,\mathrm{m^2}$, the power output could reach up to $5000\,\mathrm{kW}$ or $6800\,\mathrm{hp}$. As an annual average, a towing kite should be able to reduce fuel requirements, and consequently carbon dioxide emissions, by 10–35%. Under optimal conditions savings of up to 50% may be possible.

Modern wind-based propulsion can therefore make a considerable contribution towards reducing carbon dioxide emissions. Emissions in shipping can even be eliminated completely if biofuel or renewably produced hydrogen is used to cover the residual fuel requirements of conventional ship engines.

14.4.2 Solar Ferry on Lake Constance

Solar boats offer climate-compatible travel on water even if there is no wind, especially for short distances. The Helio solar ferry on Lake Constance has been connecting the German town of Gaienhofen to Steckborn in Switzerland since May 2000 (Figure 14.15).

The 20 m ferry can carry up to 50 passengers. Two 8 kW electric motors can provide a maximum speed of $12\,\mathrm{km\,h^{-1}}$. The boat batteries ensure a range of 60–100 km. The roof of the boat consists of an optically very successful PV system that produces an output of 4.2 kW, which provides most of the power needed for charging the batteries.

In addition to its climate-compatibility, this solar boat also offers other environmental benefits. It runs very quietly and, unlike conventional diesel engine ships, does not emit unpleasant exhaust fumes. The construction of the ship only causes minimal waves and, as a result, does not contribute towards any further erosion on the lake's shores.

Figure 14.15 The Helio solar ferry has been crossing Lake Constance since 2000. Photos: Bodensee-Solarschifffahrt, www.solarfaehre.de.

14.4.3 World Altitude Record with a Solar Aeroplane

Hot-air balloons were the first flying machines people used. Fire from firewood or straw produced the hot air needed for the carbon-neutral powering of a balloon. Today hot-air

Figure 14.16 NASA's Helios aeroplane, powered only by solar cells, set a new world record for altitude in 2001. Photos: NASA, www.dfrc.nasa.gov.

balloons are usually powered by natural gas burners. However, these balloons are highly unsuitable for freight transport or regular services. Without exception, propeller and jet-powered aeroplanes rule commercial aviation. The kerosene used is produced from oil, so the prospect of climate-friendly air travel is still a long way off (Figure 14.16).

But an unmanned light aeroplane called Helios, after the Greek sun god, showed that fossil fuels and flying do not have to be inextricably linked. The plane, developed by NASA and the California company AeroVironment, had its maiden flight in 1999. A total of 62 130 silicon solar cells with an efficiency of 19% were located on the wings, which have a span of 75.3 m and a depth of 2.4 m. These solar cells delivered the power for 14 electric motors with a total output of 21 kW. High-performance lithium batteries enable the plane to fly even after sunset.

Due to the low power output, the flight speed at low altitudes was less than $45\,\text{km}\,\text{h}^{-1}$. However, the performance of this aeroplane is not attributed to its speed but to its flying altitude. Flying over Hawaii at an altitude of 29 524 m on 13 August 2001, this plane set a world record for non-rocket-operated planes.

The maiden flight of Helios took place in 1999. Sadly, on 29 May 2003 it broke apart during a test flight and plunged into the Pacific Ocean near Hawaii.

14.4.4 Flying Around the World in a Solar Plane

The Swiss psychiatrist, scientist, and adventurer Bertrand Piccard is hoping to have more luck with his solar aeroplane. He is mainly known for his trip around the world in a hot-air balloon in 1999. With his successor project, Solar Impulse, he wanted to circumnavigate the world in a glider powered solely by solar energy. Unlike the NASA Helios aeroplane, this was a manned flight.

The project started in 2003, and a prototype was developed and tested for the first time in 2009. In 2011, construction of a second plane started, which had a pressurized cabin and other technical improvements. With a wingspan of 72 m, it is wider than a Boeing 747. It weighs just 2300 kg and is designed for an average speed of $60\text{--}90\,\text{km}\,\text{h}^{-1}$.

Figure 14.17 Bertrand Piccard of Switzerland aimed to fly the aeroplane Solar Impulse round the world in 2011. Photos: Solar Impulse/EPFL Claudio Leonardi.

The wings are equipped with 17 000 solar cells. They supply enough energy to power the aeroplane without the need for any additional energy. During the day high-performance batteries store some of the solar energy so that the plane can also fly at night.

Due to the low speed, some of world circumnavigation stages lasted several days, for example to cross the Pacific. During the day, the solar propulsion brought the aircraft to altitudes of up to 12 000 m. Powered by the solar energy stored in the batteries, the aircraft was then able to maintain an altitude of around 3000 m until the morning hours.

After numerous test flights, the round-the-world flight started in Abu Dhabi in March 2015. After the 4-day flight from Japan to Hawaii, which was over 8000 km long, the batteries were damaged by overheating. The repair took several months, so that the successful round-the-world flight could only be celebrated in Abu Dhabi in July 2016.

With technical advances in batteries and a resulting anticipated reduction in weight, similar planes could be used for longer flights with two pilots. Then even a non-stop round-the-world flight in a solar plane would be possible, and even passenger planes are conceivable in the long term (Figure 14.17).

14.4.5 Flying for Solar Kitchens

Based on current technology, it does not seem likely that solar aeroplanes will ever be able to replace large conventional planes. Even with highly efficient solar cells, the space available on the surface of the wings is not large enough to provide sufficient driving energy for planes carrying loads of several hundred tonnes. With the exception of a very limited possible use of biofuels, there are no options offering complete climate-compatible air travel. In the long term renewal production of hydrogen offers an alternative to fossil fuels.

Until then the only carbon-free alternative available is to not travel by air at all. Modern communication technologies and more and more interesting leisure activities in close proximity to where people live are helpful alternative options to travel. But the solarium around the corner is not really a substitute for a winter break in a sunny resort. It makes it even more difficult to stick to one's principles if cheap flights are offered at the same price as a ticket on the local city train.

Figure 14.18   Solar-mirror systems financed through contributions from environmental protection funds replace conventional diesel burners in large kitchens in India. Photos: atmosfair, www.atmosfair.de.

Section 3.5 of this book showed that, as an interim solution, investments in other areas can reduce as much carbon dioxide as is, for example, created by an unavoidable flight. The carbon dioxide emissions of a flight from Berlin to New York and back can be off-set for around €100. As part of its programme, the non-profit company Atmosfair offers measures designed to compensate for emissions and at the same time recommends considering videoconferencing and travel by train as alternatives (Figure 14.18).

For example, Atmosfair had large solar thermal systems installed in temples, hospitals, and schools in different projects at 18 locations in India. One of the projects is a solar kitchen for the Hindu place of pilgrimage, Sringeri Mutt. Diesel burners had been supplying the energy to prepare meals for thousands of pilgrims until modern solar mirrors replaced them as part of the project. The mirrors bundle sunlight onto a pipe and heat the water, which is then fed to the kitchen. A cleverly devised steam system ensures that the kitchens can still function even after sunset. Another aim of the project is to implement a transfer of technology to local enterprises. The systems for all 18 projects were made by an Indian manufacturer and by 2012 should be offsetting a total of around 4000 tonnes of carbon dioxide emissions from air travel.

14.5   Everything Becomes Renewable

14.5.1   A Village Becomes Independent

Energy supply throughout Germany is still heavily dependent on fossil fuels. All over Germany? No! A village populated by indomitable Brandenburgers has managed to establish an independent electricity and heat supply system that is completely based on renewable energies.

The village of Feldheim (more precisely the district of Feldheim within the town of Treuenbrietzen) with 145 inhabitants and 37 households is located about 60 km south-west of Berlin. Households, companies and the local administration have joined

Figure 14.19 Biogas plant, windfarm, and visitor centre of Germany's first energy self-sufficient village at Feldheim in Brandenburg. Photos: Förderverein des Neue Energien Forum Feldheim e.V., www.nef-feldheim.info.

forces and operate their own electricity and heat supply network, which was set up with financial support from the state of Brandenburg and the EU. The heat and electricity generated locally is transmitted directly to the consumers via the local grid. Traditional energy suppliers, and thus dependence on fossil fuels, are sidelined (Figure 14.19).

The heat demand is met by a biogas plant that can supply 560 kW of heat and simultaneously feed up to 526 kW of electricity into the public grid. In cases of particularly high heat demand, a woodchip heating system provides additional heat. A large heat store compensates for fluctuations in demand.

Feldheim also has a large windfarm consisting of 55 wind turbines with a total a capacity of 122.6 MW. Since the windfarm can supply up to 65 000 households with electricity, it feeds 99% of the electricity it generates into the public grid. The village therefore not only supplies itself, but also makes an important contribution to the energy transition in other places.

A 2.25 MW PV system on a former military site has been supplying electricity to 600 households since 2008. In 2015, a storage system based on lithium-ion batteries with a capacity of 10 500 kWh was added. The battery storage provides balancing energy for the public grid and can thus take on tasks previously performed by fossil power plants. By combining the various renewable systems and storage facilities, Feldheim is impressively demonstrating that renewable energies alone can already ensure a climate-neutral supply of heat and electricity today.

14.5.2   Hybrid Power Plant for Secure Renewable Supply

A hybrid power plant, which opened near Prenzlau in Brandenburg at the end of 2011, proved for the first time in Germany that a reliable energy supply can function with the help of wind power and hydrogen storage (Figure 14.20).

Figure 14.20 Hybrid power plant near Prenzlau. Photos: ENERTRAG/Tom Baerwald, www.enertrag.com.

Three 2 MW wind turbines feed electricity into the grid. If surpluses arise that are not needed in the grid, an electrolysis system generates hydrogen, which is stored in pressure tanks. If the wind turbines supply too little electricity, two CHP plants bridge the gap. These use the stored hydrogen or biogas from the connected biogas plant to generate electricity. An integrated hydrogen filling station enables climate-neutral refuelling of hydrogen cars.

The hybrid power plant at Prenzlau shows that a combination of fluctuating regenerative energies such as solar or wind power, in combination with storage and controllable regenerative energies such as biogas, can provide a reliable power supply. Further hybrid power plants are already planned in Germany. What is demonstrated here on a small scale can quite easily be transferred to our entire energy supply system. A sustainable and secure energy supply based on renewable energies by 2040 is therefore anything but a utopia.

14.6 Everything will Turn Out Fine

The variety of possibilities for using renewable energies described in this book, not least the examples in this chapter, have shown that it is actually unnecessary to endanger our future existence through an unrestrained use of crude oil, natural gas, coal, and nuclear energy. The effects of fossil energy on the world's climate are already noticeable. The consequences of climate change will soon be beyond our control if we do not take the bull by the horns and radically restructure our energy supply.

Renewable energies are already able to cover our energy demand in an affordable and climate-compatible manner. There is no basis for the widespread fear that the lights will go out if we cannot use oil, natural gas, coal, and nuclear energy. On the contrary, increasing the use of renewable energies will make us increasingly more independent of the conventional energy sources that are steadily becoming scarcer and more expensive. The use of renewable energies will also help to end some of the conflicts that are taking place at the moment.

Figure 14.21 The future belongs to renewable energies. By 2040 they could secure our entire energy supply in a climate-neutral manner and thus secure the livelihoods of future generations.

The question that remains is why are we still using renewable energies on such a small scale? There are many reasons for this. First of all, the wider public is not being made sufficiently aware of the possibilities offered by renewable energies. It is hoped that an increase in information sources such as this book will gradually close this gap in knowledge. On the other hand, everyone is hoping that the problem will somehow solve itself; many people do not really feel responsible or they have already lost hope. However, it is up to each and every one of us to make a contribution and to demand our politicians implement the necessary measures. If, right now, we all begin taking any opportunities to save energy and use renewable sources more seriously, we can still stop global warming and establish a truly sustainable energy supply system (Figure 14.21).

A

Appendix

A.1 Energy Units and Prefixes

The SI unit for energy is the joule (J). An equivalent derived unit is the watt second (W·s). Common energy quantities are usually measured in kilowatt hours (kWh). Table A.1 summarizes the units commonly used in the energy sector. Table A.2 shows the prefixes and abbreviations for energy units.

Table A.1 Conversion factors between different units of energy

	kJ	kcal	kWh	kg coal equivalent	kg oil equivalent	m³ natural gas
1 kJ (1 kJ = 1000 Ws)	1	0.238 8	0.000 278	0.000 034	0.000 024	0.000 032
1 kcal	4.186 8	1	0.001 163	0.000 143	0.000 1	0.000 13
1 kWh	3 600	860	1	0.123	0.086	0.113
1 kg coal equivalent	29 308	7 000	8.14	1	0.7	0.923
1 kg oil equivalent	41 868	10 000	11.63	1.428	1	1.319
m³ natural gas	31 736	7 580	8.816	1.083	0.758	1

Table A.2 Prefixes and abbreviations

Prefix	Abbrev.	Value	Prefix	Abbrev.	Value
kilo	k	10^3 (thousand)	milli	m	10^{-3} (thousandth)
mega	M	10^6 (million)	micro	μ	10^{-6} (millionth)
giga	G	10^9 (billion)	nano	n	10^{-9} (billionth)
tera	T	10^{12} (trillion)	pico	p	10^{-12} (trillionth)
peta	P	10^{15} (quadrillion)	femto	f	10^{-15} (quadrillionth)
exa	E	10^{18} (quintillion)	atto	a	10^{-18} (quintillionth)

A.2 Geographic Coordinates of Power Plants

Many power plants are so large that they can easily be seen in satellite images. The free Google Earth software allows you to display high-resolution satellite images on your home computer. The following list shows the coordinates of some interesting power plants. In Google Earth the coordinates can be entered directly in the following format:

+51° 23′ 23″, +30° 05′ 58″

This stands for 51°23′23″ North, 30°05′58″ East. Negative prefixes are used for South and West.

- https://www.google.com/earth   Google Earth homepage

Conventional power plants

	51°23′23″ N 30°05′58″ O	Decommissioned **Chernobyl** nuclear power plant; four reactor blocks, each with a capacity of 1000 MW. A catastrophic accident occurred in Block IV on 26 April 1986. Today a concrete 'sarcophagus' encapsulates the damaged reactor.
	37°25′13″ N 141°01′58″ O	Ruins of the **Fukushima** nuclear power plant; six reactor blocks with a total capacity of 4696 MW. On 11 March 2011 a serious reactor accident occurred after an earthquake and a tsunami.
	51°45′47″ N 6°19′44″ O	Fast breeder reactor at **Kalkar**, built at a cost of 3,6 billion euros but never commissioned. Today the Wunderland Kalkar leisure park is situated at the site of the former nuclear power plant.
	50°59′36″ N 6°40′05″ O	Lignite-fired power plant **Niederaußem** with a total capacity of 3627 MW, built between 1963 and 2003. With emissions amounting to 27.3 million tonnes of CO_2, in 2015 it was the power plant with the second highest CO_2 emissions in Germany.
	51°50′00″ N 14°27′30″ O	Lignite-fired power plant **Jänschwalde**; six blocks, each with a capacity of 500 MW, built between 1976 and 1989. With emissions amounting to 23.7 million tonnes of CO_2 per year, it is the power plant with the third highest CO_2 emissions in Germany.
	51°04′30″ N 6°27′00″ O	**Garzweiler II** open-cast lignite mine. Around 1.3 billion tons of lignite are scheduled to be extracted by 2045 across an area of 48 km². This requires relocation of 12 villages with a population of 7600.
	57°02′29″ N 111°42′07″ W	**Athabasca open-cast oil sand mine** near Forth McMurray in the Canadian province of Alberta. The oil is extracted with the aid of steam. The water demand in the region for this purpose is 435 billion litres per year.

PV installations

	48°08′08″ N 11°41′55″ O	2.7 MW roof-mounted PV system at the new **Messe München** trade fair site with 21 900 PV modules; installed in 1997, expanded in 2002 and 2004.
	39°49′55″ N 4°17′55″ W	1 MW ground-mounted PV installation at **Toledo** (Spain) with 7936 PV modules, built in 1994.
	52°31′29″ N 13°22′05″ O	Roof-integrated PV system at **Berlin Hauptbahnhof** (main train station). 780 PV modules with a capacity of 189 kW$_p$.
	50°00′13″ N 9°55′13″ O	**Erlasee** solar array (Bavaria, Germany) with 1464 two-axis tracking PV modules with a total capacity of 11.4 MW, built in 2006.
	51°55′41″ N 14°24′13″ O	**Lieberose** solar array (Brandenburg, Germany) with around 700 000 PV modules and a total capacity of 52.79 MW, built in 2009.
	51°34′13″ N 13°44′35″ O	**Finsterwalde** solar array (Brandenburg, Germany) with a total capacity of 80.7 MW, built in 2010/2011.
	32°58′20″ N 113°29′27″ W	**Agua Caliente** solar array (Arizona, USA) with a capacity of 247 MW in 2012, expanded to 397 MW in 2014.

Solar thermal installations and power plants

	54°51′07″ N 10°30′23″ O	Solar thermal district heating plant at **Marstal** (Denmark). Solar collectors with a total area of 17 000 m^2 supply 1450 houses with heat.
	37°05′42″ N 2°21′40″ W	European test centre **Plataforma Solar de Almería** (Spain) with a solar tower and parabolic trough test fields.
	42°29′41″ N 2°01′45″ O	Solar melting furnace in **Odeillo** (France), completed in 1970. 63 heliostats with a total area of 2835 square metres and 20 000-fold concentration achieve temperatures of almost 4000 °C.
	34°51′43″ N 116°49′41″ W	Solar thermal power plants at **Daggett** (California, USA). SEGS I with a capacity of 13.8 MW was built in 1985, SEGS II with a capacity of 30 MW was built in 1986, and a 10 MW Solar Two solar tower test plant was built in 1998.
	35°00′58″ N 117°33′40″ W	Solar thermal power plants at **Kramer Junction** (California, USA). SEGS III to SEGS VII, each with a capacity of 30 MW, built between 1987 and 1989.

	35°01′56″ N 117°20′50″ W	Solar thermal trough power plants at **Harper Lake** (California, USA). SEGS VIII and SEGS IX, each with a capacity of 80 MW, built between 1990 and 1991.
	35°48′00″ N 114°58′35″ W	Solar thermal trough power plant **Nevada Solar One** (Nevada, USA) with a capacity of 64 MW, commissioned in 2007.
	37°13′03″ N 3°03′41″ W	Solar thermal trough power plants **Andasol 1 to 3** (Spain) with a capacity of 50 MW each, commissioning between 2008 and 2011.
	37°26′30″ N 6°15′20″ W	**Plataforma Solúcar** (Spain) with solar tower power plants PS10 and PS20 with 11 and 20 MW capacity respectively, and three trough power plants with a capacity of 50 MW each.

Windfarms

	55°41′32″ N 12°40′14″ O	Offshore windfarm **Middelgrunden** (Denmark); 20 wind turbines, each with a capacity of 2 MW, built in 2001.
	29°10′17″ N 32°37′41″ O	Windfarm **Zafarana** (Egypt). In 2011 the installation comprised around 700 wind turbines with a total capacity of around 550 MW.
	32°13′48″ N 100°02′50″ W	**Horse Hollow** Wind Energy Centre near Abilene, Texas (USA). 421 wind turbines with a capacity of 735 MW, built in 2006.
	52°38′10″ N 13°25′47″ O	Berlin is at the bottom of the league when it comes to wind power in Germany. There is just one wind turbine, located at **Berlin Pankow**, built in 2008.

Hydropower plants

	25°24′27″ S 54°35′19″ W	**Itaipú** hydropower plant (Brazil/Paraguay). 20 turbines with a total capacity of 14 GW. The dam wall is 7760 m long and 196 m high.
	30°49′10″ N 111°00′00″ O	**Three Gorges Power Plant** in China, built between 1993 and 2006. 26 turbines with a total capacity of 18 200 MW.
	36°00′58″ N 114°44′16″ W	**Hoover Dam** (Nevada-Arizona, USA), built between 1931 to 1935; dam wall height 221 m; 17 turbines with a total capacity of 2074 MW.

✦	47°33′22″ N 8°02′56″ O	**Laufenburg** run-of-river plant, River Rhine, completed in 1914, electrical capacity 106 MW.
✦	47°34′12″ N 7°48′46″ O	**Rheinfelden** run-of-river plant, River Rhine. Europe's first run-of-river power plant, built in 1899. The power plant was recently refurbished and expanded.
✦	50°30′34″ N 11°01′16″ O	**Goldisthal** pumped-storage plant, Thuringia, Germany. The upper basin holds 12 million m³ of water. The power plant capacity is 1060 MW.
✦	48°37′08″ N 2°01′11″ W	**Rance** tidal power plant (Saint-Malo, France), completed in 1966; 24 turbines with a total capacity of 240 MW.

For readers of this book, all locations are also available as a download. After downloading the following file you can open it directly in Google Earth.

- www.volker-quaschning.de/downloads/EE-und-Klimaschutz.kmz

A.3 Further Reading

• www.erneuerbare-energien.de	Information on renewable energies provided by the German Federal Ministry for Economic Affairs and Energy
• www.energiefoerderung.info	Overview of funding opportunities for energy saving measures and renewable energy systems (in German)
• www.dgs.de	Website of the German Section of the International Solar Energy Society
• www.solarserver.de	Comprehensive information on solar thermal energy and photovoltaics (in German)
• www.solarwirtschaft.de/en	Information on solar thermal energy and photovoltaics provided by the German Solar Association
• www.sfv.de	Information provided by the German Association for the Promotion of Solar Power (SFV)

- www.wind-energie.de/english — Website of the German Wind Energy Association

- https://depv.de — Website of the German wood fuel and pellet association (in German)

- https://bioenergie.fnr.de/?__mstto=en — Website of the German Agency for Renewable Resources

- www.waermepumpe.de — Website of the German heat pumps association (in German)

- www.umweltbundesamt.de/en — Website of the German Environment Agency

- www.bine.info/en — Website of BINE Information Service for energy policy and energy research with a focus on energy efficiency and renewable energy, funded by the German Federal Ministry for Economic Affairs and Energy

- www.sonnenseite.com/en — Website containing comprehensive information on the environment, climate protection, and renewables

- www.volker-quaschning.de/index_e.php — Information on renewables and climate protection provided by the author

- www.youtube.com/c/VolkerQuaschning — YouTube channel of the author with videos on renewable energies and climate protection

References

[AGEB12]   *AG Energiebilanzen e.V.* (2012). Daten und Infografiken. www.ag-energiebilanzen.de.

[Arr96]   Arrhenius, S. (1896). On the influence of carbonic acid in the air upon the temperature of the ground. *The London, Edinburgh and Dublin Philosophical Magazine and Journal of Science* 5: 237–276.

[Bar04]   Bard, J., Caselitz, P., Giebhardt, J., and Peter, M. (2004). Erste Meeresströmungsturbinen-Pilotanlage vor der englischen Küste. In: Tagungsband Kassler Symposium Energie-Systemtechnik.

[BGR17]   Bundesanstalt für Geowissenschaften und Rohstoffe BGR (2017). *Energiestudie 2017*. Hannover: BGR. www.bgr.bund.de.

[BMWi17]   Bundesministerium für Wirtschaft und Technologie BMWi (2017). *Erneuerbare Energien in Zahlen*. Berlin: BMWi. www.erneuerbare-energien.de.

[BMWi18]   Bundesministerium für Wirtschaft und Technologie BMWi (2018). *Energiedaten*. Berlin: BMWi. www.bmwi.de.

[BP17]   BP (2017). *Statistical Review of World Energy*. London: BP. www.bp.com.

[BSW12]   Bundesverband Solarwirtschaft (2012). *Fahrplan Solarwärme*. Berlin: BWS. www.solarwirtschaft.de.

[BSW18]   Bundesverband Solarwirtschaft (2018). *Statistische Zahlen der deutschen Solarwirtschaft*. Berlin: BSW. www.solarwirtschaft.de.

[BWE07]   Bundesverband WindEnergie e.V. BWE/DEWI (2007). Investitions- und Betriebskosten.

[BWE11]   Bundesverband WindEnergie e.V. BWE/DEWI (2011). *Potenzial der Windenergienutzung an Land*. Berlin: BWE.

[CDI16]   Carbon Dioxide Information Analysis Center CDIAC (2016). NASA GISS Surface Temperature (GISTEMP) Analysis. http://cdiac.ornl.gov.

[DEW02]   Deutsches Windenergie-Institut GmbH DEWI (2002). Studie zur aktuellen Kostensituation 2002 der Windenergienutzung in Deutschland. Wilhelmshaven 2002.

[Dre01]   Dreier, T. and Wagner, U. (2000). Perspektiven einer Wasserstoff-Energiewirtschaft. BWK Bd. 53 (2001) Nr. 3, S. 47-54 und Bd. 52 (2000) Nr. 12, S. 41-54.

[EEA10]   European Environment Agency EEA (2010). *The European Environment – State and Outlook 2010, Understanding Climate Change*. Kopenhagen: EEA.

[Ene06]   EnergieAgentur NRW (2006). Stromcheck für Haushalte. Wuppertal.
             www.energieagentur.nrw.de.

[Enq90]   Enquete-Kommission "Vorsorge zum Schutz der Erdatmosphäre" des 11.
             Deutschen Bundestages (1990). *Schutz der Erdatmosphäre*. Bonn: Economica
             Verlag.

[EST03]   European Solar Industry Federation ESTIF (2003). *Sun in Action II*. Brüssel:
             ESTIF. www.estif.org.

[FNR06]   Fachagentur Nachwachsende Rohstoffe FNR (2006). *Biokraftstoffe, eine
             vergleichende Analyse*. Gülzow: FNR. www.fnr.de.

[Fic03]   Fichtner (2003). Die Wettbewerbsfähigkeit von großen
             Laufwasserkraftwerken im liberalisierten deutschen Strommarkt. Bericht für
             das Bundesministerium für Wirtschaft und Arbeit.

[Fle98]   Fleming, K., Johnston, P., Zwartz, D. et al. (1998). Refining the eustatic
             sea-level curve since the Last Glacial Maximum using far- and
             intermediate-field sites. *Earth and Planetary Science Letters* 163 (1–4):
             327–342. https://doi.org/10.1016/S0012-821X(98)00198-8.

[Fri07]   Fritsche, U.R. and Eberle, U. (2007). Treibhausgasemissionen durch
             Erzeugung und Verarbeitung von Lebensmitteln. Öko-Institut Darmstadt.

[Gasc05]   Gasch, R. and Twele, J. (eds.) (2005). *Windkraftanlagen*, 4e. Stuttgart:
             Teubner Verlag.

[Hul12]   Huld, T., Müller, R., and Gambardella, A. (2012). A new solar radiation
             database for estimating PV performance in Europe and Africa. *Solar Energy*
             86: 1803–1815.

[Hüt10]   Hüttenrauch, J. and Müller-Syring, G. (2010). Zumischung von Wasserstoff
             zum Erdgas. In: *energie wasser-praxis*, 68–71.

[iDMC17]   The Internal Displacement Monitoring Centre, iDMC (2017). Global Report
             on Internal Displacement. iDMC, Genf. www.internal-displacement.org.

[IEA17]   International Energy Agency, IEA (2017). *Key World Energy Statistics 2017*.
             Paris: IEA. www.iea.org.

[IEA18]   International Energy Agency, IEA (2018). *Energy Statistics 2018*. Paris: IEA.
             www.iea.org.

[IPC05]   Intergovernmental Panel on Climate Change, IPCC (2005). *Carbon Dioxide
             Capture and Storage*. New York: Cambridge University Press. www.ipcc.ch.

[IPC07]   Intergovernmental Panel on Climate Change, IPCC (2007). *Climate Change
             2007: The Physical Science Basis*. Genf: IPCC. www.ipcc.ch.

[IPC15]   Intergovernmental Panel on Climate Change, IPCC (2015). *Climate Change
             2014 Synthesis Report*. Genf: IPCC. www.ipcc.ch.

[ISE12]   Fraunhofer ISE (2012). *100 % Erneuerbare Energien für Strom und Wärme in
             Deutschland*. Freiburg: ISE. www.ise.fraunhofer.de.

[Kal03]   Kaltschmitt, M., Merten, D., Fröhlich, N., and Moritz, N. (2003).
             Energiegewinnung aus Biomasse. Externe Expertise für das
             WBGU-Hauptgutachten. www.wbgu.de/wbgu_jg2003_ex04.pdf.

[Kem07]   Kemfert, C. (2007). Klimawandel kostet die deutsche Volkswirtschaft
             Milliarden. In: Wochenbericht des DIW Berlin 11/2007, S.165-169.

[Kle93]   Kleemann, M. and Meliß, M. (1993). *Regenerative Energiequellen*. Berlin:
             Springer Verlag.

[Köni99]   König, W. (ed.) (1999). *Propyläen Technikgeschichte*. Berlin: Propyläen Verlag.

[LBEG12]   Landesamt für Bergbau, Energie und Geologie LBEG Niedersachsen (Verf.) (2012). Untertage-Gasspeicher in Deutschland. In Erdöl Erdgas Kohle 128. Jg. 2012, Heft 11, S. 412-423.

[LBEG16]   Landesamt für Bergbau, Energie und Geologie LBEG Niedersachsen (2016). Untertage-Gasspeicher in Deutschland. In Erdöl Erdgas Kohle 132. Jg. 2016, Heft 11, S. 409-417.

[Lok07]   Lokale Agenda-Gruppe 21 Energie/Umwelt in Lahr (2007). *Leistungsfähigkeit von Elektrowärmepumpen*. Lahr: Zwischenbericht.

[Mar13]   Marcott, S.A., Shakun, J.D., Clark, P.U., and Mix, A.C. (2013). A Reconstruction of Regional and Global Temperature for the Past 11,300 Years. *Science* 339: 1198. https://doi.org/10.1126/science.122802.

[Men98]   Mener, G. (1998). War die Energiewende zu Beginn des 20. Jahrhunderts möglich? In Sonnenenergie Heft 5/1998, S. 40-43.

[NOAA13]   National Climatic Data Center NOAA (2013). State of the Climate, Asheville 2013. www.ncdc.noaa.gov/sotc/.

[Qua12]   Quaschning, V. (2012). Energieaufwand zur Herstellung regenerativer Energieanlagen. www.volker-quaschning.de/datserv/kev.

[Qua13]   Quaschning, V. (2013). *Regenerative Energiesysteme*, 8e. München: Hanser Verlag.

[Qua16]   Quaschning, V. (2016). Sektorkopplung durch die Energiewende. Studie. HTW Berlin.

[Rah99]   Rahmstorf, S. (1999). Die Welt fährt Achterbahn. In: *Süddeutsche Zeitung* 3./4. Juli 1999.

[Rah04]   Rahmstorf, S. and Neu, U. (2004). *Klimawandel und CO$_2$: haben die "Skeptiker" recht?* Potsdam: Potsdam-Institut für Klimafolgenforschung. www.pik-potsdam.de.

[Sch02]   Schellschmidt, R., Hurter, S., Förster, A., and Huenges, E. (2002). Germany. In: *Atlas of Geothermal Resources in Europe* (ed. S. Hurter and R. Haenel). Luxemburg: Office for Official Publications of the EU.

[Sha12]   Shakun, J.D., Clark, P.U., He, F. et al. (2012). Global Warming Preceded by Increasing Carbon Dioxide Concentrations During the Last Deglaciation. *Nature* 484: 49–55.

[SHC17]   IEA Solar Heating and Cooling Programme SHC (2017). *Solar Heating Worldwide 2017*. Gleisdorf: IEA SHC. www.iea-shc.org.

[Sur07]   Šúri, M., Huld, T.A., Dunlop, E.D., and Ossenbrink, H.A. (2007). Potential of solar electricity generation in the European Union member states and candidate countries. *Solar Energy* 81: 1295–1305. http://re.jrc.ec.europa.eu/pvgis/.

[UBA07]   Umweltbundesamt (2007). *Stromsparen ist wichtig für den Klimaschutz*. Berlin: Umweltbundesamt.

[UBA14]   Umweltbundesamt (2014). *Schätzung der Umweltkosten in den Bereichen Energie und Verkehr*. Dessau: Umweltbundesamt.

[UBA17]   Umweltbundesamt (2017). *Übersicht zur Entwicklung der energiebedingten Emissionen und Brennstoffeinsätze in Deutschland 1990-2015*. Dessau: Umweltbundesamt.

[UNF17]   United Nations Framework Convention on Climate Change UNFCCC (2017). *National greenhouse gas inventory data for the period 1990–2015*. Doha: UNFCCC. www.unfccc.de.

[Vat06]   Vattenfall Europe AG (2006). *Klimaschutz durch Innovation – Das CO_2-freie Kraftwerk von Vattenfall*. Berlin: Vattenfall.

[Wen13]   Weniger, J. and Quaschning, V. (2013). Begrenzung der Einspeiseleistung von netzgekoppelten Photovoltaiksystemen mit Batteriespeichern. 28. Symposium Photovoltaische Solarenergie. Bad Staffelstein, 6.-8. März 2013.

Index

Renewable Energy and Climate Change, Second Edition. Volker Quaschning.
© 2020 John Wiley & Sons Ltd. Published 2020 by John Wiley & Sons Ltd.

Printed and bound by CPI Group (UK) Ltd, Croydon, CR0 4YY

16/04/2025

14658559-0001